CLINGING
TO THE EDGE

Sternula albifrons

CLINGING TO THE EDGE

A YEAR IN THE LIFE OF A LITTLE TERN COLONY

RICHARD BOON

PELAGIC PUBLISHING

First published in 2024 by
Pelagic Publishing
20–22 Wenlock Road
London N1 7GU

www.pelagicpublishing.com

Clinging to the Edge: A Year in the Life of a Little Tern Colony

https://doi.org/10.53061/ERZX7937

A CIP record for this book is available from the British Library

ISBN 978-1-78427-489-4 Hbk
ISBN 978-1-78427-490-0 ePub
ISBN 978-1-78427-491-7 PDF

Cover image: *Little Terns II* © Robert Greenhalf S.W.L.A.
All other photographs © credited persons

Typeset in Scala by S4Carlisle Publishing Services, Chennai, India

Printed by Short Run Press Ltd, Exeter

for all the watchers, wardens and volunteers who have helped ensure
the survival of Yorkshire's Little Terns

The Beacon Ponds Little Tern colony, pictured from the south in midsummer. As well as the Little Terns, also breeding here are Ringed Plovers, Oystercatchers, Avocets and Black-headed Gulls. Among other species caught in the frame are Common, Arctic and Sandwich Terns, though only Common Terns attempt to breed.

CONTENTS

PREFACE

The first time I saw Beacon Ponds was over 20 years ago. Two birder friends and I were visiting the Spurn area one mid-September. We had come to look for migrant birds (which might include the odd rarity, we hoped) along the peninsula, but things were rather quiet so we thought we'd walk back up to the Ponds, some distance to the north, to have a look at the Little Tern colony.

It wasn't there.

We hadn't realised the terns' breeding season had ended two or three weeks earlier, and that the best we could hope for was a very late bird or a straggler passing through from colonies further north.

We saw neither. Nor anything else much, as it turned out.

It was low tide, so the waders that might have been present were away feeding on the exposed Humber mud, well out of sight to the west. (We might have seen a lone and very distant Little Stint still on the Ponds – or at least something we could persuade ourselves was a lone and very distant Little Stint.) Winter wildfowl were yet to arrive in any numbers, and departing common migrants passing through on their way south were thin on the ground.

To be honest, I didn't much care for the site. The hide was ramshackle, the view from it rather bleak and uninviting – melancholic, even. There are few places more desolate than a seabird colony after the seabirds have left. It's like a stage set after the play has ended: what had been a vibrant, dynamic world, full of life, colour and noise, becomes at a stroke resoundingly empty, mundane, a dull shell of a place. The weather didn't help: it was dead still and overcast. The rural landscape I grew up in, at the southern tip of the western Pennines, embraces you in its low hills, little valleys, woods and soft, pastoral farmland;

here, you felt exposed by the wider landscape of endless arable farmland, barely above sea level, which stretches out beyond the Ponds featureless and flat down to the Humber. There was no embrace here, merely the land, the sea and the endless sky staring sullenly at each other in a kind of stubborn stand-off. It was all a bit depressing.

To be fair, the gloom may not have been entirely unconnected with our failure to find any 'good' birds that day. It was certainly not enough to stop us visiting again, several times over the coming years, though always a little too late for the Little Terns. What I came to realise, however, is that the landscape and weather in this part of the world always have an overarching influence on the visitor's psyche – more so, perhaps, than in any other place I know. Of course, that influence is by no means always so gloomy. Far from it: it can be gloriously uplifting, calmly reflective, provokingly unreliable, fascinatingly unsettling. But the key to it, to that power, is changeability.

'Spurn' – the name is generally (and rather sloppily) applied to the whole area north of the peninsula, encompassing the small village of Kilnsea and extending almost as far as the larger village of Easington – is not really one place but many places, always in flux, transformed by season, sea and weather under the great bowl of the huge sky. It's an 'edgeland', not only literally so but also in its specific, ecological sense: one of those landscapes on the edges of, and part-shaped by, human occupation, ancient and modern. (Easington is home to a charming, restored medieval tithe barn; it's also the site of one of the largest North Sea gas terminals in the country.) These are hybrid landscapes, where people and the natural world still live in constant dialogue: saltmarsh and farmland, shingle beaches and manicured caravan sites, shifting dunes and engineered floodbanks. Even the sea itself is not untouched: the horizon is dominated by a huge windfarm; beyond that are the gas and oil rigs.

Ruined military buildings from the wars are particularly prominent across the whole area, many of them fighting their last battles against the sea. For here is the other great feature of this place: the relentless conflict between sea and land, a conflict that only the sea can win in the end. The land continually shifts, mutates, forms and reforms, tries to squirm out of the way. But it's a losing battle.

I came to be beguiled by the special character of the place.

I'm using 'place' in its fullest sense here: not just location, but location with human history and culture, as well as natural history, taken into account, and the way these thicken and enrich landscape. There's a word for it, coined as early as the sixteenth century but little used today: 'chorography'. I began this book as an attempt to describe the experience of conserving one species on one site in one season, and it basically still is that. But as I wrote, I also increasingly found myself reflecting on the particular place in which all this was happening – as well as on some of the wider conservation issues that inevitably arose.

I already had some experience as a volunteer in the years before I retired, doing survey, recording, report-writing and public-engagement work at RSPB Fairburn Ings on the border between West and North Yorkshire (another literal and metaphorical 'edgeland', brilliantly formed from the reclamation and remodelling of a landscape scraped to the bone by open-cast coalmining). So when my wife and I moved to our village just north of Spurn, I quickly made myself available for several volunteering roles at the Spurn Bird Observatory – the 'Obs'. The main one, however, came in an unexpected form: an invitation to chair the management committee (or 'steering group' – there's a bit of an issue about nomenclature around these parts: the names of things shift uncertainly, see below) for the Little Tern colony, a project led by the Obs. The group comprised representatives of the various bodies involved in supporting the project. My protests that I knew nothing about Little Terns fell on deaf ears: it seemed I didn't have to. My past life as what was flatteringly referred to as 'a senior academic' meant I knew how to chair meetings (apparently); everyone else in the group would have the requisite knowledge and experience. Further protests – that my background was in the humanities not the sciences (not even a Biology O-Level) and that chairing meetings was one of the things I very much did not miss about work – were airily dismissed.

I caved.

So my first priority was to build some understanding of the birds and the place. Answers were to be found in the project's annual reports, which dated back to the early 2000s. These seemed to be complete, but weren't always easy to locate. And they appeared under a variety of titles – nomenclature again – such as 'Easington Little Tern Protection Scheme Breeding Report', 'The Little Tern Breeding Season at Beacon Lagoons Nature Reserve'… 'Beacon Ponds Nature Reserve'… Were these all the same place? And the naming of parts went beyond the reports and into the field. What's the difference between a lagoon and a pond? And why 'Ponds', not 'Pond'? I could see only one body of water around which the terns bred, though there were more… 'lagoons', apparently, just to the north. Did the terns once breed there? And if so, why not now?

Further investigation only added to the confusion. Where was 'Long Bank' and was it the same as 'New Bank'? Some folk called what I thought must be New Bank, Flood Bank. Others called the whole lot Long Bank. They are all floodbanks, but… And – see above – where does 'Spurn' actually stop and start? Was it the same as the Observatory's Bird Recording Area or not? Trying to sort out who owned or leased various bits of the area, and which government agencies had oversight and legal responsibilities, was another source of bewilderment. One thing the various reports had in common was rows of logos along the bottom of the title page identifying partners and funders – yet these changed often but irregularly, some disappearing, others popping up in their place.

There was a lot to go at. The answers to some of these questions appear in this book, but it's sensible to sort out a few basics here.

'Easington Lagoons' historically refers to the whole complex of pools (sorry) and marshes behind the line of dunes which separates them from the beach and the North Sea. Beacon Ponds is the southernmost of these and the area to which the terns are now restricted.

'Ponds' rather than 'Pond' because while for most of the year it is one body, falling levels of water in the late summer and autumn expose islets, spits and ridges that to a greater or lesser extent split the area into two, three and sometimes even four, separated though interlinked cells.

As for the 'Long Bank' minefield, I refer the reader to the map on page xvii. I've consulted the Ordnance Survey Map (Explorer 292 Withernsea and Spurn Head) – which cuts through the Gordian knot of confusion.

'Spurn' (never mind its recording area) I take to refer to the peninsula, starting at its northern end, where the land narrows dramatically as the North Sea and the Humber Estuary close in on each other. Anywhere north of that belongs to Kilnsea, where the peninsula starts to fatten out, then, five-and-a-half kilometres further on, Easington.

The reports contained much that was useful about the terns, but naturally assumed a basic knowledge I didn't have. Remedying that was my next task.

So, I quickly learned that the Little Tern is one of a genus of the seven smallest species of tern, *Sternula* (lit. 'little tern'), which are common around the world, from North and South America to Eurasia, South Asia, Africa and Australia.[1] They are similar in appearance, habits and behaviour. Two – Saunders's Tern (*Sternula saundersi*) and 'our' Little Tern (*S. albifrons*, lit. 'white forehead') breed in the Western Palearctic, but separately, only coming close to each other in Arabia. (The North American Least Tern [*S. antillarum*, lit. 'of the Antilles', where it winters] is an extremely rare visitor to northwestern Europe, though is so similar to *S. albifrons* that it is likely to be under-recorded.) Little Tern is the smallest. It's about the size of a Song Thrush, but more slender, longer-winged and extraordinarily light in the air (it weighs no more than a tennis ball). It is among the rarest of Britain's breeding seabirds. A summer migrant here, it winters on the coasts of West Africa, arriving in late April to breed along our shores. It needs sandy and shingly beaches, though will also use, as in our case, the margins of adjacent saline lagoons. It returns south at the end of August. Colonies are concentrated along our southern and southeastern coasts, though there are also important sites in Scotland, Wales and Ireland. The colony at Beacon Ponds is the only one in Yorkshire.

Although it is generally doing reasonably well throughout its range, the Little Tern has seen severe declines in the British Isles over the last decades, largely because its vulnerability to natural threats has been significantly increased and exacerbated by the impact we humans have had and continue to have on its ecology. Loss of and disturbance to habitat in particular have meant reductions in populations and so greater susceptibility to predation, bad weather and other natural threats. Thanks to a concentrated conservation effort, however, its population has stabilised with something in the region of 1,400–1,500 pairs now breeding here. Still, only the Roseate Tern, an occasional passage visitor to Spurn, is rarer among the British terns.

So far, so good. The basics were in place.

Five years on, I've learned a lot, as I hope this book testifies. Not least that the basics are just that – basic, and that there is much we still don't fully

---

1 Advances in genetics continue to refine – we might say complicate, or even confuse – how we ascribe species taxonomically. The little tern species were originally lumped together with the larger tern family, *Sterna*. And recent research proposes a new subspecies of *albifrons*, the 'Levant Little Tern' (*S. a. levanticus*) to add to those already identified.

understand or even plain know. Some of what we *do* know might not be that certain, either. Nor do the basics give a sense of the particular character of this long-distance traveller: small, fragile, vulnerable, yet having the strength and resourcefulness to hang on in our nature-depleted and too-often unsympathetic country. In many ways, the Little Tern could stand for any number of other species, and not just among the birds.

So, this book is not a how-to manual, though I hope it contains things that might be of practical use to wardens and to volunteers both actual and potential. Nor is it a scientific treatise: if there are any new behavioural or other insights, they are entirely accidental. It's more an as-it-happened account, month by month, of the ups and downs, the pleasures and anxieties, of one Little Tern colony in one particular year – 2022 – and one particular place, Beacon Ponds, on the edge of South Holderness, East Yorkshire. It's essentially the story of one person's involvement at the sharp end of conservation. I've tried to avoid as much as possible allowing the wisdom of hindsight to creep into the monthly accounts. Much of it was written, in my head or occasionally on my phone, while I was on shift helping monitor the colony: it seemed important to try to capture the lived experience of the work. That also opened up the opportunity for more general reflection: speculation about the birds' behaviour, about the nature of our relationship – good and bad – with them, about what they mean to us. At times this has prompted larger questions, some personal, some more general, about conservation, about the way it works… and sometimes doesn't.

I hope you enjoy it.

Richard Boon, December 2023

ACKNOWLEDGEMENTS

The 2022 Team

Head Warden: James Wilson

Assistant Wardens: Rob Hunton, Mick Turton

Residential volunteers: Tom Wright, Murphy Hand, Ben Secker

SBOT volunteers: Zach Pannifer, Bethany McGuire

YWT interns: Claire Galpin, Harry Appleyard

Local volunteers: David Constantine (SHCS), Brian Walker, Paul French, Georgia French, Marcus Brew, Richard Boon

Occasional and visiting volunteers: Paul Collins and Rob Adams (SBOT), Mike Pilsworth (RSPB), Mike Coverdale, Leslie Ball, Angela and John Taylor

Management Committee

Chair: Richard Boon

Members: James Wilson, David Constantine (SHCS), Mike Pilsworth (RSPB), Rosie Jaques (YWT), Laura Carmichael and Andrea Treasure (EA) and Emily Paterson (NE).

Project Consultant: Mike Coverdale. Jackson Sage (HNP) also provided support and advice.

It has become something of an 'Acknowledgements' cliché to thank those who have saved the author from error, while also stressing that any errors which remain are the author's own. Cliché perhaps, but never more true than here.

My greatest debts are to our wardens; James Wilson, whose commitment and engagement raised the bar on what a warden can achieve and without whose meticulous and comprehensive observations and record-keeping this book would have been a great deal harder to write; Rob Hunton, no less committed and thorough, and who has generously read and commented on drafts; and Mick Turton, whose participation may have been curtailed in the 2022 season but whose contribution to the Little Tern Project overall has been invaluable.

Mike Coverdale, former SBOT and Little Tern Project Management Committee Chair has unparalleled knowledge of the history of the colony and the Project and has been as generous as ever in sharing it: he also read and commented on parts of the draft. Mike Pilsworth of the RSPB has likewise been an invaluable source of knowledge and advice. Dave Constantine of SHCS has been most helpful in sharing his personal collection of documents relating to SHCS and has generously provided the majority of the photographs used in the book. Dr Kieran Lawrence provided valuable input on ringing, Tim Jones, Director of SBOT, on egg-collecting. Paul Collins, SBOT Head Warden, read and advised on my comments about ringing and migration. The 2023 wardening team of Jacob Spinks, Tom Wright and Harry Coghill have kept up the good work, supported by Eva Finney, Erin Gold and Joe Griffin.

Jonnie Fisk, Toby Phelps, Lucy Mortlock, Sarah Harris, Chantal MacLeod-Nolan and Liam Andrews also deserve my thanks.

All the photographs in this book have been taken by locals associated closely with the Little Tern Project and the Obs: David Constantine and Thomas Willoughby in particular, and Tom Wright, Rob Hunton, Mike Pilsworth, Jackson Sage, Georgia French and Bethan Clyne. I am grateful to each of them.

Thank you to Nigel Massen and all at Pelagic, not least for their faith in an author with a very slight track record of natural history publishing.

Last, and very far indeed from least, I thank my wife Rebecca for her steadfast and loving support. I'll decorate the snug now.

ABBREVIATIONS

BOU	British Ornithological Union
BTO	British Trust for Ornithology
DEFRA	Department for Environment, Food and Rural Affairs
EA	Environment Agency
ERYC	East Riding of Yorkshire Council
GPS	Global Positioning System
HNP	Humber Nature Partnership
NE	Natural England
PSPO	Public Space Protection Order
RSPB	Royal Society for the Protection of Birds
SAC	Special Area of Conservation
SBO(T)	Spurn Bird Observatory (Trust)
SHCS	South Holderness Countryside Society
SPA	Special Protection Area
SSSI	Site of Special Scientific Interest
YWT	Yorkshire Wildlife Trust
YNU	Yorkshire Naturalists' Union

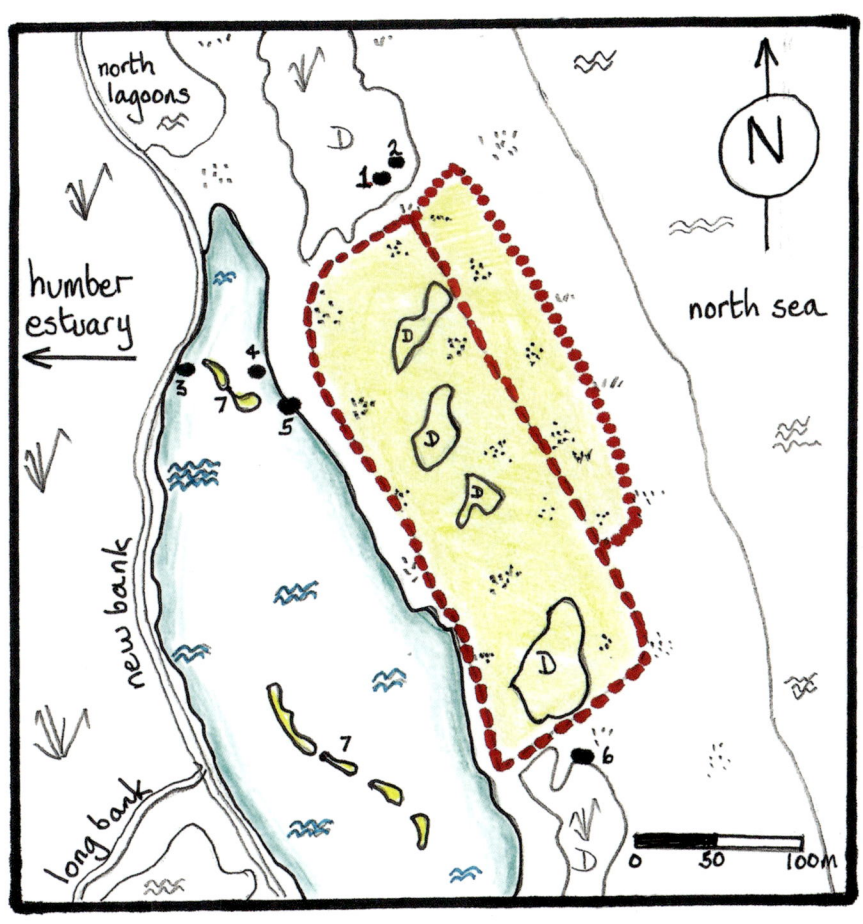

BEACON PONDS 2022

1. old hut
2. beach viewpoint
3. viewpoint
4. 'tern table'
5. beached raft
6. new hut
7. low-water isles

· · · · · · · · · · · electric fence
· · · · · · · · · · · beach extension
D. large dunes

Beacon Ponds seen from the air in the spring, looking south-west. The Humber Estuary is at the top of the picture, the North Sea at the bottom. Kilnsea Wetlands is to the left, the southern tip of Easington Lagoons to the right. The Little Tern colony sits centrally along the eastern side of the Ponds behind the dune line.

INTRODUCTION

At first the beach seems unprepossessing, grey-brown, lifeless, flat.

It's largely devoid of vegetation, made up of shoals of sand, fine and coarse gravels, shingle, pebbles, cobbles and a few larger rocks. Walk over it, over the soft, crisp crunch of the finer gravels – always easiest to walk on, apart from the damp sand – and you realise that it's seldom as flat as it seems. There are ripples, dips, bowls and ridges, some quite high, and all to be shifted, sorted and remodelled by the next tide. And look closer, especially where the falling water has left the shore still wet, and the shingle – let's call it that, all of it, for ease of reference – takes on a richer texture. Shades of grey and brown dominate, it's true, but much is painted in different colours: reds, blacks, whites, creams, orange-yellows and purples even, often striped, veined, blotched or speckled. There are bits of seaweed, too, scattered thinly along the strandline (seldom a lot: there's little for it to get a hold on in these seas), as well as relics of past lives: crab shells, cuttle bones, razorshells, piddock-drilled rocks, bits of gull-scavenged fish and bird corpse, driftwood branches bleached white by sea and sun. From time to time there will be a rotting Seal or Porpoise carcass. And lives from the far deeper past as well: there are fossils to be found here, not so many as there are on Yorkshire's Jurassic Coast 60 kilometres to the north, but still not too difficult to uncover… ammonites, belemnites, sponges, and the splendidly named Devil's Toenails (*Gryphaea*). This is Yorkshire's Cretaceous Coast.

(There's a large human component to the beach, too. But we'll come to that later.)

We are at the southern end of Bridlington Bay, on the Holderness Coast, just north of Spurn Point. You don't often hear 'Holderness Coast' without the prefix 'fast-eroding'. 'Fastest in Europe' in fact, as those of who live on it tire of hearing. The shore erodes quickly because it is composed of soft boulder clay, laid down 12,000 years ago under retreating glaciers, and it is from the detritus left behind that our shingle beach has largely been created. There are

The beach in winter, looking south towards Kilnsea.

granites from the Lake District, Scotland and Scandinavia and larvikite from Norway; there are sedimentary sandstones and fossil-bearing limestones too, the latter ranging from dark grey cobbles, sleek as a seal's head, to brilliant white chalk – the same chalk that forms the cliffs of Flamborough Head at the north end of the bay but which lies deep beneath the boulder clay here. There are flints, pieces of coal, even Whitby jet if you are lucky, and metamorphic rocks which tell of ancient volcanic violence: split the dullest cobble and it may reveal sparkling iron pyrites or quartz. The sand itself, mineral-made, glitters in the low sun.

This is the kind of beach beloved of Oystercatchers, Ringed Plovers... and Little Terns, all of which will, given half a chance, nest on it above the high-tide mark. (Usually, anyway. Sometimes they are caught out by a spring tide, especially when it is storm-driven; sometimes, they just get it wrong.) It's a harsh and difficult place to breed: not only are nest-scrapes open to tide and weather; they are, once found, an easy target for both land and aerial predators. So all these birds rely on camouflaged eggs and nestlings, on precocial youngsters that from a few days old can sprint for cover like battery-driven toys when threatened, and on parents who will defend them by aggressive mobbing or distraction.

There are many beaches like this around the British Isles, but few of them, unless they are protected, will hold sizeable populations of these nesters. Ringed Plovers are in decline, Oystercatchers too. Little Tern numbers, though they remain relatively healthy elsewhere in their range, have been falling for decades in the British Isles: they peaked at 2,800 in the late 1960s and early 1970s, but are now half that. They are our second most at-risk seabird (after their congener, the Roseate Tern) and have been Amber-listed since 2015.[2] And although new colonies have developed, often with our assistance, the overall picture is one of loss, of their range around the British Isles becoming ever more restricted.

The culprit, I'm afraid, is us.

Human disturbance, whether in the form of infrastructure development – buildings, roads, sea defences, marinas – or leisure activity, has cost the Little Terns dear. Their favourite places are also among ours: to sunbathe, fish, play, beachcomb, walk the dog, fly the kite. And the loss of any colony puts increasing pressure on those that remain. So the colony at Beacon Ponds, though small, is important.

Whether Little Terns ever bred in great numbers on this beach, we can't be sure, though it is believed some at least did so a little to the south. In fact, there is not a great deal that we do know about the history of their presence in the British Isles: terns are present in the fossil record but can be difficult to separate confidently into different species. The Little Tern was first recorded in Britain in 1671, but had clearly been present for much longer. (As a species it was first described by the great all-round natural historian Simon Peter Pallas in 1764, though its original taxonomic status has since become ever more modified as scientific method has become increasingly sophisticated.) It is not, I think, an overly optimistic assumption that the birds would have been a

---

2 The UK Birds of Conservation Concern Lists chart the decline of bird species and so help prioritise conservation planning. As of 2021, the Red List identifies the most at risk (70), and the Amber, those of moderate concern (103). Green-listing indicates those species that seem to be at least stable, though species can decline into Red and Amber with alarming speed. Both the Red and Amber lists are growing.

A newly hatched Little Tern.

regular and familiar presence on the Holderness beaches for hundreds of years at least, and quite possibly for thousands. They still breed here, and have, on and off, for several years, but in small numbers and with varying success, even when protected. Their main colony now is not on the beach but immediately adjacent and separated from it by a line of 'grey' – that is, established, stable and vegetated – sand dunes. They nest in the sand and shingle on the shore of the small brackish lagoon that is Beacon Ponds, the southernmost of the larger complex of pools and marshland that is Easington Lagoons.

Beneath the substrate on which the terns and other shore-nesters breed is, perhaps surprisingly, old agricultural land. Indeed, farming, mainly of cereals and rape, continues only a little further inland, and agriculture of one kind or another has always dominated the area. Archaeological surveys have revealed medieval and Romano-British field systems, and activity still earlier into prehistory: there are the remains of a burial mound just to the north of the Ponds, and a few years ago a henge was briefly exposed by erosion on the edge of the beach before itself succumbing. But when those Romans farmed, they could stretch another five kilometres out to the east, such have been the historical losses to the sea. No fewer than 23 historical settlements that we know of have gone under the waves along the Holderness coast. And it seems very likely that our terns would have nested along beaches now lying as gravel beds on the sea floor.

The recorded history of Yorkshire's Little Terns begins, however, not at Beacon Ponds, nor even on its adjacent beach, but a few kilometres to the south on the Spurn Peninsula.

In 1861, between 40 and 50 pairs were reported as breeding but, in a pattern that was to be repeated for decades to come, breeding was disrupted and eggs plundered by day-trippers, mainly ferried in from across the Humber, and local children.

In fact, the 1860s represented something of a peak in egg-collecting and shooting activities, with the cliffs of Bempton and Flamborough Head, across the bay to the north, being particular blackspots: one contemporary estimated that over 200,000 birds were killed or their eggs taken in a single season. Birds were shot from the cliffs or from boats, and eggs taken by collectors, or for the leather-patenting industry, or for restaurants. (It is remarkable, at least to me, that some top-end restaurants continue to offer Black-headed Gull eggs on their menus, at £5.50–8.00 each. Fortnum & Mason regards the arrival of gull eggs – "responsibly sourced by licensed collectors" – as a highlight of the culinary year. The eggs are from Yorkshire.) Again, contemporary accounts are keen to suggest that the problem lay mainly with 'cheap trippers' rather than local 'climmers' (climbers) who depended on the birds for their living. Opposition to the slaughter came from a partnership between those we would now call conservationists and upper-class shooters who wanted their pastime kept exclusive – an uneasy alliance that was to prove to have historical longevity. What also motivated many of these early conservationists was Christian faith and their belief (made all the more urgent by the fate of the Great Auk) that life was sacred, so it is unsurprising that two of the prime movers of the seabird campaign belonged to that great tradition of the Victorian parson-naturalist. The Rev Henry Frederick Barnes-Lawrence, who initiated the campaign, was a vicar in Bridlington, while the Rev Francis Orpen Morris (of whom a little more later), also based for most of his life in a Yorkshire parish, published remarkably widely in natural history, including the eight-volume *A History of British Birds* (while remaining a diehard anti-Darwinist). Also involved were Alfred Newton, the foremost professional ornithologist of his day and founder of the British Ornithological Union (BOU) and John Cordeaux, leading

John Cordeaux

amateur and a figure well known to historians of the Spurn area, not least as author of *Birds of the Humber District* (1872), and in particular for his pioneering work on bird migration.

These are some of the key figures in the history of bird conservation in Britain, playing major parts in the creation of organisations such as the RSPB. Also involved in the Yorkshire initiative were a surgeon, a poet and an MP, and in 1868 the East Riding Association for the Protection of Seabirds was formed. The following year the MP, Christopher Sykes, introduced a bill to Parliament which became the 1869 Seabirds Preservation Act: the first piece of legislation to defend seabirds, though not, at the pleas of the 'climmers', their eggs. Even so, in 1874 no fewer than 89 shooters had to be prevented from going on to the cliffs in the Flamborough area: as was to prove the case many times in the future, it is one thing to legislate, and quite another to enforce that legislation. Progress on protecting birds' eggs remained a painfully slow business. Not until 1950 was general protection made law (if again difficult to enforce), and not until 2000 were penalties extended to include imprisonment – a sanction that has proved genuinely effective.

In his *Birds of the Humber District*, written three years after the ground-breaking legislation, Cordeaux wrote of the Little Tern (or, as he knew it at the time, the 'Lesser' Tern):

> This beautiful and delicate-looking little creature, the smallest of our English Terns, nests annually at Spurn Point, but in greatly reduced numbers, the Act for the Protection of Sea-Fowl coming just in time to prevent its complete extinction.

The pressures on the Spurn birds, as elsewhere, were primarily shooting – wings and feathers were supplied to the millinery trade in London – and egg-taking. As we have already seen, the 1869 Act didn't protect birds' eggs, so even after the shooting stopped, or at least focussed on wildfowl, the problem for Spurn's Little Terns remained 'eggers'.

Writing in *The Naturalist* a few years later, Cordeaux notes that

> about a mile north of the Lighthouse there is a colony of the Lesser Tern, which unfortunately is becoming smaller each year through the senseless and relentless plunder of the nests by excursionists.

Help arrived in the form of a County-Council-appointed 'watcher' in 1895, but it tells us something about just how relentless the plunder was that when the post was briefly terminated, the issue immediately resurged, with, for example, one visitor from Grimsby reported in 1898 to have taken 34 eggs. Another watcher was employed in 1900 and 200 young were successfully hatched that season.

Despite the appointment of further watchers, now the responsibility of the Yorkshire Naturalists Union (YNU), to safeguard the interests of the terns (and Ringed Plovers), their foothold on the peninsula remained insecure, and their numbers at best modest, up to the Second World War:

in 1936 there were only 94 nests in the whole area. (It is interesting to note, however, that in 1912 80 nests were found in an unspecified location – almost certainly on the beach – north of Kilnsea, with only scattered breeding to the south.)

The two World Wars of course had a huge impact: the whole area was heavily militarised to counter the threat from across the North Sea. The Spurn Peninsula, guarding as it does the mouth of the Humber, was particularly affected. Influxes of large numbers of personnel and the establishment of gun placements, tank traps, pill boxes, barracks and other military buildings, a railway line and later a road changed the character of the place for ever, and in ways still evident today, 80 years on.

In particular, the new road made visitor access to the whole area so much easier in the postwar years as cars became more affordable and popular. Under the YNU's guardianship, the terns held on until the late 1950s, but

The remains of military buildings are visible across the whole area.
Here, a gun emplacement from the Second World War battery at Kilnsea clings on.

their productivity was low. In 1953, 53 pairs laid 101 eggs, but only one chick managed to fledge. That was also the year of the great winter tidal surge and floods that stretched the length of the east coast and took over 500 lives. Defensive measures were subsequently put in place locally with the creation of a flood bank (New Bank) and the establishment of a lagoon, which effectively if unintentionally began the history of the Easington Lagoons/Beacon Ponds site.

It's worth pausing here for a brief overview. The first point is that Little Terns have always been close to the edge in Yorkshire, and not just literally: their hold on their breeding sites has been tenuous, their numbers fluctuating but never more than modest, their breeding success limited. They have flirted with local extinction. The second is that they have survived largely through the dedication and hard work of mainly amateur naturalists, with support from the authorities and other organisations that has not always been consistent or reliable. Progress, in terms of legal protections, has been painfully slow and seldom entirely adequate. Despite that, the merest glance at the situation in the 1860s is enough to show the enormity of the change in attitudes towards nature generally that has developed over the last 160 years. It is now impossible to imagine the free-for-all slaughter on Bempton Cliffs or the often mindless plundering of eggs on Spurn Peninsula, so fundamental has been the shift in public attitudes. It is not of course complete, nor even, perhaps, entirely secure: the continuing history of conservation here and beyond still faces many of the same problems encountered by the early pioneers, as we shall see. But it is important to remember – sometimes simply to hang on to – the distance already travelled.

The modern history of the Yorkshire Little Terns effectively begins in the 1970s. The late 1950s and the early 1960s were the closest that the species came to total failure, with only a few nests along the peninsula and low success rates. From 1964 to 1976 there was scarcely a nest to be found; a pair attempted to breed in 1977 but the nest was washed out. In the same year, however, a second pond at the Easington Lagoons site was created when the tide broke through into the borrow pit excavated when raising sea-defences. It is this lagoon which has come to be known as Beacon Ponds, after the historical warning beacon nearby that was lost to the sea. (Beacon Lane, the approach to the area from Kilnsea in the south, is another commemoration.) What had been developed was the only area of shingle, sand dune, swamp and saline lagoon in the region, good habitat for breeding shorebirds, including Little Terns – though the colony did not form there until the early 1990s – and, in terms of specialised invertebrates and plants, a special ecological niche. A Southampton University survey of coastal saline ponds in England and Wales in 1989 concluded that the site, though relatively new, was home to "a number of species [which] seems to be increasing steadily, and with appropriate management, the site is well worth conserving". The potential was confirmed by a 2015 report to Natural England by the University of Hull Institute of Estuarine and Coastal Studies, which found that the Ponds contained such specialists as the rare flowering plant Spiral Tasselweed, Green Hairweed and other algae, various marine worms, shrimps and snails, as well as Sand Gobies and Nine-spined Sticklebacks. Our own observations have added European Eels among others to the list.

Coastal lagoon habitats are, however, variable and subject to relentless change. Coastal defences might solve one set of problems, but they create others (often at considerable distance from the area protected). Flood banks inhibit and prevent natural change, and sediment from erosion brought in by tide and storm builds sand dunes, so squeezing the area in between. (It is this process of sedimentary build-up that over millennia has built, shaped and reshaped the Spurn Peninsula, and continues to do so, producing what was once memorably described by the Yorkshire ornithologist John Mather as a very slowly beckoning finger of land between the North Sea and the Humber.) What might have 'gone down' to saltmarsh at Easington Lagoons remains as brackish ponds of variable salinity, refreshed by the sea through storms and high tides, and perhaps by percolation through the dunes; at the end of a long dry summer the water can be hypersaline. Winter rainfall dilutes it again.

The result of this coastal squeeze has been that Beacon Ponds itself has shrunk by almost one third overall since the turn of this century, and the Little Tern colony has edged progressively southward. In years gone by the colony was spread more to the north and formed two or three smaller breeding areas, presenting particular challenges for a succession of wardens. The compression and incremental move towards an area of thickly vegetated marsh at the southern end will effectively stop the terns dead in their tracks and eventually lead to the abandonment of the colony (something to which we will return later).

Tern numbers were initially slow to build at the colony, but simple numbers of breeding birds aren't the whole story. In 1984, only eight pairs bred, but they raised 23 young: a productivity rate of 2.88 chicks per pair which any Little

The colony site looking north. Beacon Ponds are to the left, the North Sea to the right. Easington Gas Terminal is just visible in the far distance.

Tern warden in the country would kill for. The YNU had resumed its role as the overseeing body in 1980 and a local birder, Arthur Piggott, voluntarily wardened the area until 1988. For the next six years the site was supported by a new initiative, the Spurn Heritage Coast Project, a body formed by representatives of the Countryside Commission, the Nature Conservancy Council, Humberside County Council, Holderness Borough Council, Easington Parish Council and the Yorkshire Wildlife Trust (YWT). Then came the stewardship of the South Holderness Countryside Society (SHCS).

Formed in 1982–83, the role of the society has been absolutely central in creating and supporting the Easington site, and it remains actively involved to this day. Not least among its achievements has been the purchase and leasing of land in the area, crucial in protecting the landscape and its ecology. Land was purchased from a local farmer in 1989, then more leased from the Environment Agency (EA), which had taken over control from the Yorkshire Water Authority, before still more was bought in 1991 and 2000. The following year, responsibility for wardening the site was taken over by the Spurn Bird Observatory Trust (SBOT, created in 1946 by the YNU), based just over a kilometre to the south, which has continued as the lead partner in the enterprise to the present.

The period 1986–2002 saw Beacon Ponds hosting 1.5–2 percent of the British population of Little Terns, with a productivity of 0.44 fledglings per pair – close to the national median (at least as represented by those colonies monitored). The subsequent decade saw the beginnings of a movement from simple protection and monitoring to a more active engagement in trying to improve breeding success and thus increase numbers. Under the leadership of SBOT, some limited experimentation was tried. For instance, natural refuges in which chicks might hide from bad weather or predators were supplemented by artificial ones: these simple shelters were made of materials found in the area, such as rocks and driftwood, but also washed-up plastic rubbish, especially short sections of agricultural drainage pipe. Results were variable, but shelters made from natural materials appeared to be the more attractive, and helped a number of chicks survive that might otherwise not have done. Decoy birds, designed to encourage more of the real thing, were tried, though without the accompanying soundscape of calls their effectiveness was limited. Lion dung (from Longleat, I believe) was put down to deter Foxes – but failed to do so.

In 2002, 34 pairs fledged 32 young (a productivity of 0.94 fledglings per nest); in 2005, 27 fledged 28 (1.01); in 2011, 25 pairs produced 19 (0.76). These are good productivity rates, but sadly were not representative of the decade overall, for which the average number of fledglings produced was only around a dozen per year (off a minimum of 23 and a maximum of 45 pairs). The main problems, as usual, were predation, bad weather and a combination of the two – with human disturbance thrown in for good measure. In 2003, 16 nests were wiped out in 36 hours by Crows, and Crows were also largely responsible for destroying 21 nests, containing 47 eggs, in 2009. Foxes, too, were a recurrent problem. Three successive years of bad weather in the middle of the decade had a major impact; a colony weakened by wet, cold and rough weather is more open to attack. Hence, 2006 saw a productivity of 0.08, with Foxes capitalising on particularly bad weather. Crows, ironically, were not a big problem in this season – but they were the next, when the attentions of 22 of them, along with more bad weather, saw 27 pairs of terns produce only three young, none of which fledged. Productivity: zero. Crows did for 2009, Foxes for 2010. On top of this, there was also a recurrent problem with vandalism, the worst example of which was in 2003 when the windows and wall panels of a hut were repeatedly smashed in and abuse and outright aggression directed at the warden.

Most seasons also saw the terns displaying over other, traditional sites along the peninsula. In 2003 two nests were predated at the Point, while three at Chalk Bank were lost to high tides. High winds buried nest sites under sand at Chalk Bank in 2008, pushing birds to the Point and then, possibly, over the Humber to the north Lincolnshire coast. The following year, the same year Crows wiped out the Beacon Ponds colony, and the terns relocated to the peninsula (there were counts of over 80 at Chalk Bank) but later returned: some re-laid at the Ponds, but then simply deserted, for unknown reasons, abandoning their eggs in early July. If ever there were a season that showed how tenuous a grasp on the area Little Terns can have, and how weather and predator can conspire against them, it is 2009.

What was becoming increasingly evident was that proper safeguarding and monitoring of the colony was beyond the capacities of a single warden, hardworking and committed as they all had been. Under the leadership of Mike Coverdale of SBOT, 2011 saw something of a management step-change. A partnership project was established, with greater funding found for a second, assistant, warden starting a month after the first – as breeding activity reaches its busiest stage. Electric fencing had been used since 1990, but now was upgraded from poultry netting to the more robust and (notionally at least) reliable multi-stranded fence. In 2013 lead warden Heather Bennett was able to introduce night-time wardening for the first time, enabled not only by the presence of the second warden but also by an increasing involvement of volunteers; she also wrote what remains an invaluable handbook for Little Tern monitoring and protection. And in a side-development, an MSc student from Leeds University spent the season researching her dissertation on factors contributing to the reproductive success of the Little Tern; with some missing years, the arrangement with Leeds has continued to the present as part of a commitment to incorporating some serious scientific research into the project.[3]

---

3 Charlotte Davies's work looked at the behaviour and provisioning of chicks throughout June and July. It was enabled by special permission from the BTO (under the Schedule 1 licence held by the SBOT warden) and involved temporarily – and harmlessly – marking chicks on their backs with coloured pen in order to identify individuals and family groups. The marking was lost with the down as feathers grew. One key finding was that, contrary to earlier research, it appeared that hatching success was lowest in those pairs that laid earliest. See Bibliography.

The benefits of the enhanced approach were quickly apparent. In 2011, 25 pairs of the terns fledged 19 chicks (0.76); in 2012, heavy predation (mainly by Fox) didn't stop 11 chicks fledging from 23 nests (0.49); in 2013, 36 pairs produced no fewer than 40 fledglings (1.11), while 2014 topped that with 60 from 45 (1.33). There were, of course, poor seasons too: nobody involved in any kind of conservation work expects success to be guaranteed, and reasons for failure aren't always clear. Of 25 pairs in 2016 only six bred, fledging five chicks (0.83); this might have been down to poor weather early on, but the birds have triumphed over bad starts in other years. Fourteen fledglings at 0.29 in 2017 was partly the result of high tides washing out 17 of 49 nests (the highest number of nests since 2006, possibly due to relocations from a failed colony to the north at Crimdon Dene near Hartlepool), but unresolved problems with the electric fence saw Fox account for 50 percent of the other eggs and chicks. Short-eared Owl may have played a part too. It was a similar story in 2018, when Fox was again largely responsible for only four chicks fledging from 33 pairs (0.12). An Oystercatcher even got in on the act, being seen to take three eggs. Meanwhile, the terns' levels of interest along the peninsula remained low and never developed beyond the merely speculative. While these kinds of failure are at the time undoubtedly demoralising for all concerned, it's important to hold on to the long view… and maybe have a little blind faith.

The following two seasons saw a productivity level of 1.56 off a total of 81 successfully fledged chicks.

What benefited the project for much of the decade was its participation in a major EU initiative, the 'Life+ Project', which focused on developing our understanding of the lives of Little Terns and the problems they face, with the overall aim of helping them recover their numbers. Given that the years between 1969 and 2002 saw a 25 percent decline in the UK population on top of previous losses, and 2000–2015 a further 18 percent, the renewed focus, which was conceived on a national level (over 20 sites participated) and prioritised the development of knowledge of the birds' biology and habitat needs, and sharing it, came not a moment too soon. The EU provided half the funding, the rest being found by participating partners, who included county councils, the Wildlife Trusts, the RSPB and the National Trust. And although the project concluded in 2019, the RSPB has continued to lead a 'Life on the Edge' initiative, dedicated to pursuing the same goals and working to a Recovery Plan published in 2018. The loss of funding has been significant, but retaining the ambition and clarity of the plan is crucial, as is sustaining the momentum it created. In our case, thanks to some indefatigable work by our fundraisers, we have managed to raise enough money to employ, since 2021, a third warden: one charged with a particular focus on public engagement.

Time, I think, for another overview.

Many different organisations are involved in conservation projects, not only at national (and international) level, but also locally. Some are professional, many voluntary, some governmental, others charitable, some new and short-lived, others older but operating under new names (with remits that might differ enormously or barely at all), some with statutory powers, others with none. Overall, the key change in conservation over the last few decades has been its professionalisation. As Cabot and Nisbet point out in their authoritative New Naturalist Library study *Terns*, this has brought many benefits, not the least of which is the centrality to the work of trained scientists with expertise in a necessary variety of related fields who can develop coherent strategies on national and international levels (though it's also worth noting here the rise of citizen science, the value of which resides primarily in large-scale data-gathering

Welcome signage at the southern approach to the colony, with partners' logos.

*Why we ask visitors to stay near the tideline and keep their
dogs on leads: spot the Ringed Plover chick...*

but which is increasingly finding real expertise among amateurs). The less
happy outcome is the ever-increasing bureaucracy involved: "Anyone who tries
to understand why, how or by whom tern populations are managed is quickly
submerged in an alphabet soup of acronyms."

So... today Easington Lagoons are a designated Site of Special Scientific Interest
(SSSI), as is the nearby Humber Estuary. The two together make up a Special
Protection Area (SPA). The Estuary is also a Special Area of Conservation
(SAC), but the Lagoons aren't. (Do keep up.) The SSSIs are designated so by
the UK government and have regulatory force. The wider area also has international status as a RAMSAR site – not an acronym, but named for the city in Iran
where it was signed in 1971, and properly known as the 'Ramsar Convention
on Wetlands of International Importance Especially Waterfowl Habitat' (try
getting a viable acronym out of that one). All such designations recognise and
enshrine the value of the sites they cover but none, many feel, has real legal
teeth, not even the SSSI. (While the UK was a member of the EU, the SPA
would – we think, we're not entirely sure – have taken precedence over the
SSSIs.) Underpinning protection from the law originates in the Wildlife and

Countryside Act of 1981 (and its subsequent amendments), the key element of which for us is that it makes it illegal "to intentionally or recklessly disturb at, on or near an 'active' nest". Even here there are issues of efficient policing. As for our own 'disturbance', whether it be predator control, bird ringing or habitat management, consents and licences must be obtained from the relevant authorities. Complex mosaics of land ownership and leasing can mean lengthy applications for consents before works may be undertaken, while permission to take action against an unforeseen predation issue may take too long to process by overworked government agency staff to be effective.

And then, of course, there's money.

Participating partners in local projects may all contribute, sometimes in kind rather than hard cash, but have many calls on what are often small budgets anyway. Grants have to be sought from funders large and small, public and private, regional and national (sometimes international). Each has its own deadlines and qualification requirements and by no means all are able to, or wish to, provide funds over a long period of time. The politics of conservation, too, shift: some years ago we were funded, generously, by a huge multinational company. Most conservation organisations now find that company's continuing commitment to fossil fuels makes it unacceptable as a donor. (Our most recent large donor has been a North Sea windfarm company, whose donations are distributed through a professional community-funding company.) And conservation itself slides up and down successive governments' lists of priorities.

In the ordinary course of things, many applications must be made. Many fail. In the case of Beacon Ponds, I can find between 15 and 20 different funders over the last 50 years or so, and I'm confident I haven't traced every last one. We are of course genuinely deeply grateful to all of them, but…

These are the realities of modern conservation. Negotiating this world takes huge amounts of time and energy, and is only possible through the dedication of those involved, very few of whom, professional or voluntary, are anything other than hardworking and dedicated, whether they be overstretched government agency employees, reserve staff or volunteers. Time that for some at least might be spent better in the field is too often whiled away in meetings or in front of computer screens. What might be achieved if all that energy and commitment could be focused into one, all-embracing, properly resourced, consistent effort?

So this is what we look forward to at the start of the 2022 season: a head warden, supported by two assistants and three residential volunteers, backed up by a group of local and visiting 'vols' who can ease the pressure on the main team, provide last-minute cover and plug unpredictable gaps in the wardening rota; a management committee representing all the main partners in the project (SBOT, RSPB, YWT, SHCS, NE and EA) to which the head warden reports at monthly meetings and to which outside guests with particular experience or skills may be invited to contribute. Clear protocols are followed on everything from monitoring to data collection and sharing, to ringing, to public engagement to – inevitably – health and safety. Our funding is secure for the season. We have no visiting research student this season, but the arrangement stands and we will continue to benefit. We are part of a national group from which we gain valuable knowledge and experience and to which we contribute our own. And because ours is a small colony, our set-up enables an intensity of scrutiny, monitoring and protection that the larger colonies simply cannot manage to the same degree.

The operation is in as good a place as it has ever been.

All we need now are the birds.

THE 2022 SEASON

APRIL

The season is upon us.

I'm uneasy. I've been re-reading Richard Mabey's memoir *Nature Cure*. It's about his experience as the victim of the blackest of depressions and how he seeks renewal in and from the natural world. But he's plagued by anxieties about the slow return of our summer birds, which are starting to migrate in from Africa and southern Europe – or should be. My own brush with mental illness is (I hope) far behind me, and nature has for me too been a vital part of the cure. But at this time of year I'm always reminded of Mabey's irrational fretfulness: he knows the birds will return, but… *what if they don't?* What would that mean?

The earliest ever record of a Little Tern arriving in the wider Spurn area is 5 April (in 1886), but now they typically arrive at Beacon Ponds in the last two weeks or so of the month; sometimes a little later, seldom much earlier. By the 11th of this month birds are being reported from the south coast, from Kent, Hampshire and Dorset. A few have been seen at various inland reservoirs, too, and one has reached as far north as Tiree.

But none at the Ponds.

Other shore-nesting birds, which share the Ponds site with the terns, have on the other hand already started to nest. Resident Ringed Plovers, themselves under increasing threat (like the terns, largely through habitat loss and disturbance), began displaying in early February.

They will soon need our protection too: their main predators at this time of year, Carrion Crows, have been on territory since January and are also well into their breeding cycle: plover eggs will be easily available and they pack a useful nutritional punch for hungry nestlings. Last year, a Little Owl also took at least a couple of plovers, and a bird was heard calling at a nearby farm earlier this year.

If our resident birds start early, then so do we. Our preparations began in Christmas week, in fact, when a small team of us gathered to clear Sea-buckthorn from the southern end of the site. We walked in from the north, and it was immediately apparent that the Ponds had had other visitors before us. The lagoon immediately to the north (long abandoned by the Little Terns as its habitat became less suitable, but still within the SSSI), was fringed around its entire circumference by orange-dyed corn grain – duck bait for wildfowlers, whose empty cartridges also littered the place. It was a ludicrous amount of bait that would only pollute the water. It is not uncommon for nature reserves to abut shooting reserves (RSPB Titchwell is a good example). Indeed, some wetland nature reserves began as, and were managed for, wildfowlers, and while their interest in the birdlife is clearly differently motivated, it is quite possible for good working relationships to be forged between them and conservationists, on a pragmatic if not necessarily an ideological level. In our case – and again, this is not untypical – an arrangement was made when SHCS took control of the land years ago that shooting could continue on pools to the north, with the Ponds itself being left alone. Nonetheless, this incident felt very much against the spirit, if not the letter, of the agreement.

We walk on, down through the empty colony. Sea-buckthorn is a spiny shrub which dominates much of the peninsula to the south. In fact, it is common along our eastern coasts and in places has been deliberately introduced to help bind sand dunes as defence against the ever-encroaching North Sea. Its dense structure provides cover for tired migrant Scandinavian thrushes in the autumn, and its orange berries a useful refuelling resource. It can also be a haven for great rarities – warblers from Siberia, buntings from the far north, pipits from Central Asia, pioneering or simply lost birds – and is one of the key sites in the UK for finding them. But the shrub is very invasive and its thicket-forming habit quickly comes to dominate the landscape, effectively forming a monoculture. It is by no means so dominant around our site but left to its own devices it soon would be, and while it is not the only unwelcome colonising plant we have to deal with, it is one of the most problematic. Removing Sea-buckthorn is not an easy or pleasant job. As you cut the bushes down and remove them, spines find ways to penetrate your skin however careful you are and however thick your protective clothing (last year one of our team found one slowly working itself out of his arm months later). The remaining stumps are then treated with a particularly unpleasant herbicide. We never like resorting to chemical treatments, but sometimes it is unavoidable – and, of course, used only under licence. And the work is essential: Ringed Plovers might tolerate some encroachment of vegetation on to their breeding space, but Little Terns hardly ever do. Unless we keep the colony clear, we will lose it. It is a thought to be clung too as another insidious spine sneaks through and the December wind finds the gaps in your clothes.

Our next preparatory visit came in mid-January: a beach-clean.

First up, a damaged, beached raft. For a number of years we have experimented with floating a couple of wooden rafts on the Ponds to provide further secure nesting habitat for the Little Terns. We spread sand, shingle, larger pebbles and sometimes smashed seashells on the surfaces to encourage the birds, and put in a grid of battens to stop the whole lot sliding to one end in windy weather.

It has been an emphatic failure.

So far as I know, in the last five or six years only two Little Terns have actually landed on the things, let alone nested on them; they have shown virtually no interest at all. On the other hand, Black-headed Gulls and Avocets *have* (though not always productively), and there is value in that, particularly regarding the Avocets. Not only do they need our help and protection too (they are another Schedule 1 listed species) but, being highly alert and aggressive defenders of their territories, they offer a useful early-warning system when predators appear and provide another line of defence for shore-nesters generally. So the rafts have been worth persisting with, even though they are expensive, difficult to maintain and prone to being seriously damaged by the winter weather (it is surprising how destructive wind and wave can be even on the sheltered Ponds). One has given up the ghost this year, and needs to be dismantled and removed; the others, including the recently installed and experimental small platform raised a metre above the water on legs and wittily if not entirely accurately christened our 'tern table', will do.

Next, we do a quick check on the 'blue-rope fence' (the outer of our fences, designed to keep the public from wandering too close to the colony). It will need to be mended and, in one or two places, excavated: it is a sign of how much the winter seas change the profile of the beach that fenceposts last seen standing 1.2 metres high now protrude by half that, with the lower of the two lines of ropes they carry buried under the sand for tens of metres along the length of the fence line. Elsewhere, posts have now 'grown' in height as sand has been removed, and have become unstable and in need of re-fixing. All this can happen literally overnight. There's little point doing much now, though: there's plenty of winter weather left to undo any repairs we might make. A job for the calmer days ahead.

Sand can be a problem in other ways, too: driven by high winds during the breeding season, it can cover the eggs of both the Little Terns and the Ringed Plovers. Both species have been known to remove it to save their nests, but not always. Wind-blown sand has other, more general impacts as well: as it accumulates in the colony it gives a footing for plants to get hold, and will slowly but relentlessly add to the growing dune system. Again, there's little we can do now, but how best to manage our habitat is always on our minds and is becoming ever more of an issue. Fortunately, the sea is on our side.

A couple of weeks before there had been a 'wash-through', where a high tide, driven higher by winds from the northern quarter, had broken through gaps in the dune system over the beach and into the Ponds. (This was the second time this winter: a rather bigger event occurred last November.) Later in the year and it would be a disaster for the colony (it has happened before, in 1991 for instance); in the winter months it is by no means a bad thing, reinvigorating the brackish waters of the Ponds, sweeping away smaller, nascent dunes and shaving the edges of the larger ones, killing off even some of the hardier colonising plants, and, with luck, adding another layer of pebbly substrate to the breeding areas. Such events, however (it would happen again in February), sometimes exact a price, and not just affecting the integrity of our fence: they sometimes sweep the beach clean of accumulated rubbish, but they can also bring it all back, with extra added. This is what our team gathers to collect and remove.

The Ponds in winter can be a rather dismal place. Under a heavy, rain-threatening January sky they can be positively dreary. If there is a sea-fret on (known locally as a 'rork'), the whole landscape takes on the greyscale colours of the winter gulls. The wet mist dissolves the boundaries of sea, sky and land, blurs the hard edges of things. And though there may be literally hundreds of wintering birds present, with flotillas of ducks loafing about and, at high tide, waders from the Humber waiting for the tide to drop and their muddy feeding-grounds to be exposed once more, the site can seem oddly lifeless. Those grey,

A minor wash-through after a high tide and gales. This is a small, stream-like incursion; such events generally occur on a wider front and a much larger – and sometimes very violent – scale, refreshing the brackish waters of the Ponds.

massed ranks of snoozing Knot and Dunlin, heads tucked under, look like little more than cobble fields, extruded from the shingle and the dark wet sand. There is colour present, but it is muted, in the modesty of the winter-plumaged Teal, Wigeon and Little Grebe… apart from the plastic.

A single traffic cone lay on the beach, its brilliant dayglo orange a metaphorical poke in the eye. I wish I could say its presence was unusual, and it might have been once; now, it is depressingly unsurprising. The amount and variety of human rubbish, much of it plastic and similarly lurid, which can appear on the beach and wash through into the colony, is astonishing. It is dominated by fishermen's waste: nets and lines; blue, red and green crates; crab and lobster pots, storm-ripped from their moorings along with the empty white plastic drums used as their marker buoys; and rope, ranging from three-strand orange nylon to hemp hawsers as thick as your forearm, in large, seaweed-laced knots and tangles rooted immovably in the sand.

There are always fragments, too, both massive and small, of old sea and miliary defences in wood, concrete and brick, as well as detritus from the land: tyres, from tractors, cars and bikes; sheets of plastic, torn and frayed; carpet and other assorted textiles; rusted old engine parts; drainage piping, domestic and agricultural; bottles, of course, glass and (overwhelmingly) plastic; hard hats; children's balloons; cigarette lighters; footballs; and once (noteworthy

Storm-driven high tides deposit rubbish all over the beach,
especially along the strandline, and on to the colony site.

even for our battle-hardened volunteers) a near-intact sofa.[4] It's not just us, of course: it's all our North Sea neighbours. We have identified rubbish from the Low Countries, from the Baltic, from France. Some objects, sea-battered and scoured, are simply undecipherable. Much of the smaller stuff blends in surprisingly well among the shingle, driftwood and bits of torn seaweed, and most of the larger items are easily collected and removed. Exceptions – the genuinely and immediately hazardous – are uncommon, but unnerving when present: an open, and badly resealed metal drum labelled 'Sulphamic Acid' (a de-scaler and rust-remover used on boats' hulls) was clearly not to be trifled with. But all needs to be dealt with: even those harmless things that might fit aesthetically, as it were (the lobster pots, lengths of driftwood) still provide structures around which sand accumulates, contributing to the growth of sand dunes and the development of the 'coastal squeeze'.

And this is only the big, obvious stuff. Oil seems much less of a problem ever since tankers exiting the Humber were banned from flushing their tanks out in the 1970s, but the number and regularity of smaller, less dramatic spillages from oil rigs remains a big issue. Coal dust from defunct coastal collieries miles to the north regularly etches the sand as black strandlines when the tide falls. In itself, this is probably relatively harmless, but it reminds us of the other pollutants in our seas, pollutants invisible to us and far beyond our immediate control: organophosphates, microplastics, heavy metals and other chemicals, floating in suspension or stirred up by dredging. Much comes from agricultural and industrial discharge from the land, a phenomenon that began in earnest with the Industrial Revolution and has done nothing but worsen since. Algal blooms and mass mortality events of, for example, seals, auks and (probably) crustaceans,[5] make the damage dramatically visible, but mostly the problems at any one moment are beyond our senses and so easily forgotten.

---

4 Among the more outré items to have appeared are a perfect miniature plastic toilet, false teeth (bottom set only), a partly dismembered bicycle and a giant inflatable doughnut. One year wardens started a collection of disembodied doll parts which became a rather unsettling presence in our hut.

5 As I write in 2023, there is a continuing crisis in the North East of a mass mortality of crabs and lobsters which awaits satisfactory explanation, with chemical poisoning, algal bloom and an as yet unknown disease all in the frame.

And with the threatened abandonment of so much environmental protection after Brexit, the problem can only grow. Sewage discharge into rivers and thence into the sea has become a national scandal. Again much of the resulting damage is not immediately visible (though try telling that to our wardens as they pick sanitary wear off the fences after a storm) but it is there and it affects – pollutes, poisons – all life in the ocean, from microscopic life-forms in the seabed sediment all the way up the food chain to our terns, fishing for the sand eels that will feed them and their chicks.

Our modest team can't clean that up.

The sea is a rubbish tip, and the tides vomit a tiny proportion of that rubbish back on to beaches. Yet even the rubbish that can't be removed helps cast a light of the nature of this wild place. I'm reminded of the work of the Yorkshire-based Alice Fox, who spent 2012 as YWT Spurn Point Artist-in-Residence, making art – textiles, drawings, videos, pieces of writing – out of the natural phenomena of wind, water and sand and the way they interact with the human-made world. She notes the play of the wind on water, the patterns of ripples on the beach, the way the hard leaf-tips of Marram-grass, eddied by the wind, will describe perfect circles in the sand. But she also sees and incorporates the rubbish. Sea, sand and weather are her collaborators, and the art she makes with them is a shared process, using found, collected and arranged materials – whatever they are. Shells, feathers, pebbles, a cotton reel, a bit of seaweed, a plastic fork, an unidentifiable bit of vivid plastic are collected and displayed together in interplays of colour, shape and texture. She ties lengths of fabric around groynes and allows the sea to dye, stain and mark them as it will, before removing them, drying and sewing them all together to hang in the Spurn lighthouse or to return to the beach for the incoming tide to push and bend into new, sinuous shapes. Fox's art speaks not just to the natural, but to the historical and cultural; like the elements themselves, it is a great homogeniser, and it makes our thoughtless discards its own.[6]

Take driftwood: it is difficult, sometimes, walking the beach, to distinguish the branch of a tree from a fragment of shattered wood piling, both washed in from

---

6 The delicate and complex relationship between humans and nature in this part of the world is also explored in Tom Wells's play *Big Big Sky*, which opened at the Hampstead Theatre in London in 2021. Wells, a Kilnsea-born playwright, makes one of his protagonists a Little Tern warden.

Alice Fox, Spurn Cloth (2012)

lord-knows-where-and-when, after the sea's relentless buckling, abrading and polishing of them. Massive, reinforced concrete blocks, remnants of sea- and military defences (hah!) are smashed to smithereens, the detritus then sculpted into shapes and colours all but identical to the rocks and pebbles around them. A large section of brickwork, the size of a boulder and precisely the orange-red of the summer plumage of those winter Knot still dozing listlessly by the Ponds, is undeniably artificial, but the sea has eroded and distorted it into a surreal version of itself: a reflection in a fairground mirror. Conversely, that lump of igneous rock may be ancient in origin but its unpolished jaggedness betrays its most recent home and the human hands that put it there: the sea has smashed it off new defences at Withernsea to the north, constructed from imported boulders of Scandinavian granite. Single bricks mix with jagged pieces of flint and soft lumps of chalk. Even when the apparently natural origin of a 'beach cobble' is betrayed by a streak of rust where once a bolt was sunk, then that too becomes part of a unique landscape where modern industrial archaeology sits easily alongside the flints of the Neolithic and the ammonites, belemnites and crinoids of the Cretaceous. And who knows? That suspiciously symmetrical rock might once have been the hearth of a house in one of those villages long ago lost to the sea.

Or it might not.

And that is the point: this part of the coast is an edgeland, yes, but it is also an 'in-between' land, where familiar distinctions are blurred. A wild location, certainly, sometimes very wild indeed, but not a *wilderness* in the truest sense: there is too much of the human, historical and modern. There's a name for it: 'homolith', that ever-growing layer of human-made stuff, big and small, that we are spreading over the surface of the planet. It's a key characteristic of the Anthropocene. But here the sea and the weather fight back, making the whole Spurn area a contingent place: always shifting, always uncertain, made of centuries of the interaction of humans and nature, whether on the scale of the roads, farms, flood banks, borrow pits and dykes of the larger landscape or the flotsam and jetsam of the shoreline.

Such are the thoughts that can occur as you carry a traffic cone to the back of a pick-up.

It's mid-April now. Reports of terns arriving are coming in quite regularly from the London reservoirs, and as the month progresses birds are noted with increasing regularity from Nottinghamshire, Gloucestershire, Greater Manchester, Northamptonshire, Somerset, Staffordshire, Cambridge…

Still none at the Ponds.

Many of the reports, in fact, come from inland sites and all, with the possible exception of that early Tiree bird, refer to passage migrants. Little Terns are traditionally uncommon away from the coast in the UK: sightings are sporadic and generally of single birds, often seen with Arctic Terns and, if you are lucky, the rarer Black Tern. Although fossil tern bones from as long ago as just after the last Ice Age have been found well inland, they are few and hard to identify to species level. Common Tern is perhaps the likeliest candidate. It is uncertain whether Little Terns have ever bred inland as they always have, and still do, on shingle spits and islands on the Continent. (Francis Orpen Morris in the 1850s talks about the species not only being common on the Norfolk coast – still a stronghold – but also about its distribution all around Europe, including inland 'by the borders of rivers'; further back still, the great French naturalist François Belon [b.1517, murdered 1564] notes their presence around the Black and Caspian Seas, but also on the River Irtish in Siberia.) I have myself watched breeding Little Terns on the Loire in France. We cannot know for sure whether they have ever bred along UK rivers: I can find no reference to them having done so, though it is at least worth noting that Francis Willughby's 1676 *Ornithologie… of Middleton in the County of Warwickshire* includes the 'Lesser Sea Swallow' in its listing. The Yorkshire-based Orpen Morris does mention 'specimens procured' near Huddersfield, Sheffield and Ecclesfield and 'not infrequently near Barnsley, though occurring at uncertain intervals', but it is most likely he is referring only to passage birds.[7] Whatever may have been the case in the past, the one UK inland site where Little Terns breed today is on an old Second World War airfield on Tiree, where taxi-ways and runways have been smashed up and dressed with sand and gravel to create appropriate habitat; carefully managed and technically inland, but self-evidently not at all far from the sea.

Our birds are probably being held up, as they were last year, by cold winds from the northeast quarter, which have been slowing all bird migration.

We might not have a Little Tern, but we do now have a Little Tern Head Warden.

James is in his mid-twenties, softly spoken, with a dry sense of humour, and very hairy. He has been a volunteer warden on North Ronaldsay in Orkney and on the Isle of May, where he worked with Arctic Terns, and in Minnesota, where he worked with bears. He has a strong academic background, with an MSci which included a dissertation on the complexities of Robin song. Along the way he has picked up some management experience and worked with the public as a nature guide and interpreter. On top of this, and frankly rather annoyingly, he is also a talented artist, photographer and guitarist.

---

7 Orpen Morris, it must be said, is not seen as the most reliable of sources: he could seldom resist a good anecdote, however unlikely, especially if he perceived useful Christian allegory to be found in it.

And he wasn't even a shoo-in. I'm seldom less than amazed by the quality of some of the applicants we get for what is, after all, a fixed-term seasonal contract paid at national minimum wage, with long hours and heavy responsibilities ranging from the most basic of practical skills to complex data-gathering and interpretation and public engagement. In the past we have 'sold' the job as an opportunity for young naturalists, starting out, to build their experience and skills: an apprenticeship of a kind. It has certainly played its part in career development; past wardens have gone on into permanent positions, some of them quite senior, with big conservation organisations: the RSPB, the WWT and the Wildlife Trusts. Now some at least of those whom we appoint are already qualified enough to be further up the career ladder. Their time with us is certainly not wasted, and we benefit from their knowledge and experience enormously, not least in the scientific rigour they increasingly bring to our data-gathering and analysis. But I do occasionally find myself thinking they might be an under-used asset in the broader scheme of things.

James's Little Tern contract doesn't begin until 1st May, but he has begun early to help out with some more general duties as an assistant warden for the Obs. Nonetheless, he's still finding time to get a head start with his tern duties. He's quick to recruit three long-term, residential volunteer wardens to support what will be our three paid wardens (James himself and two assistants who will start later in the season), and to reassure himself that the half-dozen or so local volunteers (myself included) who usually help out are still willing and able to do their bit this year. Gear has to be taken out of storage at the Obs and transported up to the colony with the help of our colleagues at Spurn YWT; their continuing help throughout the season will be invaluable. And there is plenty of stuff to bring up: the components of the second, seasonal hide we put up at the south end of the colony and the tools to put it up with; signs to go along the colony borders to advertise its presence and ask the public not to come too close; rope and new posts to repair the fence along which the signs will be located; and most important – and most cumbersome and complicated – the components of the electric fence that will surround the colony to guard its occupants against ground predators. Setting all this up takes time, but fortunately needn't be done all at once. Crucially, the fence needs to be ready to be switched on by the time the terns, when they finally get here, are ready to settle. By the beginning of the last week of the month, James and his small team have finished repairing the blue-rope perimeter fence and erected most of the electric fence.

The northern end of the site, showing the electric fence.
The crates bottom left protect the batteries and chargers.

Much of our work over the summer will consist not just of monitoring the terns but also of protecting them against their potential predators, which are legion: gulls, corvids, owls and raptors threaten from the air, while on the ground there are Foxes, Badgers, Hedgehogs, Otters, Rats, Weasels, Stoats, Grass Snakes…

…sometimes it seems anything and everything is a threat. Not all of these have proved to be a problem at the Ponds, though all have at least been seen in the area, and if one lesson has been learned over the years it is that you can't necessarily predict what might become an issue. Another is that while you might win a fight against particular predators, you never really win the war. (And even if you could, magically, remove all predators from the site, then all you will have done is create an uncontested space for their replacements to move into.) Some of the tactics we have tried in the past have been – shall we say? – somewhat experimental. These include that lion dung, loud sonic alarms, fake 'predator eyes' (twin lights mounted in a battery-powered unit bright enough and far apart enough to persuade our nocturnal predators that bigger nocturnal predators are around). None has met with notable success. The simple, visible presence of wardens (at night aided by torches) is always effective, though even here predators learn to modify their tactics, but our most important defence is our electric fencing.

Looking back over the last 30-odd years, it is clear that electric fencing, in its various manifestations, has always been a time-consuming, difficult and at times maddening job to install, repair and keep running efficiently. Before the colony's slide to the south gathered pace, it was spread over a larger area, and in some years required three separate fenced enclosures to protect it, originally with electrified poultry netting. Even now, when the colony is more condensed and needs only one large enclosure, maintaining it at peak efficiency is a relentless and sometimes frustrating task.

It is equally clear that it has always been worth it: it works.

For the last few years the colony site has occupied approximately 24,000 square metres – or just under six acres. It is roughly rectangular in shape, with the long sides running northwest–southeast, parallel to the sea. Not all of it is usable by the terns: the broken wall of dunes – dominated by three particularly big ones – marks the seaward edge, with the greater part of breeding habitat on the landward side but with gaps and 'bays' within the dunes giving some extension. So the whole area needs to be enclosed by the electric fence. A properly maintained fence dramatically reduces, though seldom entirely prevents, the incursion of larger predators – notably, Fox, one of the biggest threats to Little Tern colonies. (In 1997, for example, 14 pairs of Little Terns which nested outside the fence lost all their eggs and chicks to what was probably one Fox.) It also makes it much harder for Roe Deer and, to a lesser extent, Hare, to get in in their search for an under-exploited plant food source; here the risk is of trampling on nests and their contents. The signs advertising its presence encourage dog-walkers passing along the beach to keep their animals under control, too.

Over the last few seasons we have used a nine-strand fence. That means there are approximately 4.5 kilometres of wire to install (if this seems a lot, bear in mind we used to use a 12-strand system…). There's a further half-kilometre of 'trip wire' to run around just outside the perimeter to deter Foxes and Badgers from tunnelling under the fence proper. And then there's the two-strand blue-rope perimeter fence, set well outside the electrified bits, to keep the public back. Later in the season we'll also need a length or two of electrified poultry netting to extend the protection down to the Ponds' waterline as it recedes. To install, run, test and maintain all this, we need: four car batteries for power, with two large metal boxes to enclose and protect them; associated chargers, testers and energisers; a fence voltage reader and its batteries; around a dozen large spools of wire plus spares; wooden strainer posts to hold the whole thing up and lightweight metal posts to carry the wires between the strainers, together with ratchets to tighten the wires and numberless insulators to stop the current from simply running to earth; 20-odd 'Beware Electric Fence!' signs to hang on said fence; anti-predator spikes to stop raptors perching on the posts we've so handily provided; four A-frame signs to set up well down the beach asking the public to stay back near the tideline; seven Little Tern information boards and nine plain 'Keep Back' signs.

Then there's the setting up. The protocols for safe and effective usage (we follow those developed by the RSPB) are more complicated and require more precise application than might be assumed. Badger and Otter, for example, need only a five-strand fence, a third of a metre high; Foxes, better jumpers by far, need nine, to a minimum height of 1.25 metres. But those nine strands are not set equidistantly; at the simplest, they need to be closer together in the lower half, so that not only Foxes but Badgers and Otters can't pass between. But the protocols go further, specifying differing gaps to cover differing contingencies. The very bottom strand can be particularly tricky: too high, and predators or clod-hopping Hares can squeeze under; too low, and there is a real threat to larger chicks running round. And beaches, of course, are not level and even things. While the fence can be bent round bigger humps and lumps, smaller ones need to be flattened to ensure clearance. Hollows can be levelled-off with large stones. Of course, there's still nothing to stop smaller predators – Weasels and Stoats, Rats, Hedgehogs, even mice – getting in underneath; at the other end of the scale, a 1.25m fence is no real barrier to the athletic and heavy-footed Roe Deer. And, to be honest, the fence doesn't always stop the determined and clever Fox, who will happily play the long game, returning persistently to test for weaknesses and willing, perhaps, to risk the odd electric smack.

The biggest potential weakness is of course losing power. Batteries and chargers might develop faults, connections work loose, but it is the wires themselves that can pose the greater problem: wind-blown sand can close the bottom gaps to the first and even second strands, allowing the electricity to run to earth. Strong winds blow in not only all manner of human rubbish, but also natural 'rubbish' in the form of matted clumps of seaweed, which seem to knot themselves on to the strands with irritating tenacity. All needs to be removed. Those same winds, when easterly, douse the fence in saltwater spray, again leading to a drop in voltage. (There is a chemical spray to correct this, but we prefer to wait for the rain to clean the wires, which it does efficiently.) Posts and wires might simply snap. So the fence needs checking several times a day, both visually and by using a hand-held electronic tester.

Installing the electric fence. The North Hut is in the background.

Electric fencing is nothing if not high maintenance.

As I've said, there's no doubting the importance of the fence to the continuing wellbeing, quite possibly the very existence, of the colony. Yet I sometimes find myself looking at it with a kind of sadness. As a symbol of what conservation can mean, what it *is*, it seems, in the face of the magnitude of problems the natural world faces, a little desperate: a tiny fenced enclosure in the wild, which (though it benefits others) is basically dedicated to the preservation of just one fragile species. It all seems so… inadequate. So last minute.

But I do the only thing we can when such thoughts occur: banish them and get on with it. And hope.

The 2022 Little Tern Protection Scheme is all but ready to go.

There's just the one thing missing…

MAY

'They've made it again,
Which means the globe's still working...'

Ted Hughes, 'Swifts'

Little Terns. Four of them.

On the first of the month, late in the day, as the sun goes down.

(The first few Swifts and the first terns usually arrive at about the same time, and so it is this year.)

James is especially pleased: after all, it's his first day proper on contract. He does well not to look smug.

These four might not even be necessarily 'ours', I suppose: they could fly on to colonies farther to the north, to Teesside or beyond, or even back south, to sites across the Humber in North Lincolnshire. They might come back in a few days or, if breeding attempts elsewhere fail, to try again here much later. Generally speaking, however, it's believed they go straight to their breeding sites, with little dawdling along the way and no 'staging' flocks. Even if they are 'ours', they'll tend not to hang around for extended periods: to do so might be to attract predators before they've even begun to breed. Safer to start that process elsewhere, perhaps: although there is a little half-hearted displaying here, with males dangling fish in front of females. One theory suggests the process proper may take place over beaches in areas where the birds have no intention of breeding. As for so much with this species, we just don't know for certain. But frankly, at the moment I don't care much: *they're back*.

Over the next week or so numbers fluctuate but overall rise slowly: three on the 2nd, five on the 3rd, eight on the 4th, then into the teens by the end of the week.

First arrivals. The bird to the left has already caught a Sand Eel.
These fish form a significant component of the terns' diet.

Our preparations need completing. We are helped by the arrival of Rob, formerly of the motor trade but now committed to a career in conservation and fresh from working for the RSPB as a monitoring and protection officer in the Forest of Bowland. He's also a very experienced 'Spurn regular' who knows the area well. Like James, his initial responsibilities are with the Obs and he will begin his formal duties as assistant warden for the Little Tern Project only at the start of next month; also like James he nonetheless gets stuck in immediately when he has the time.

The electric fence is completed, the southern, seasonal hut erected, and the northern hut repaired. The latter, known as the Old Hut, has been in place, remarkably, since 1991 – though it's been moved on a couple of occasions and the number of repairs and replacements that have been made raises the question of whether it's actually the same hut or its child or even grandchild.[8] It's prone to leaking, its door won't shut properly, it creaks alarmingly in high

---

8 Fans of the classic BBC sitcom *Only Fools and Horses* may recall the philosophical conundrum that was Trigger's broom.

winds and (especially for those of us of generous proportions) it is deeply uncomfortable. We are all very fond of it.

The southern or 'New' Hut is a more recent addition and a necessary response to the southward drift of the colony. As with the Old Hut, it's positioned so that not only can we see as much of the colony as possible, but so we can also keep an eye on what's happening on the beach. But we need to be able to see the whole site at once in order to monitor the birds properly, and that has become increasingly difficult, especially from the Old Hut, over the last few years. Hence, we have also set up what we rather grandly call a viewpoint (actually a post with one of our signs on it and a plank to sit on) directly across the water on the New Bank gabions, where we can sit in increasing discomfort and use telescopes to count and map the birds and their territories on the opposite shore.

The terns are our main but not our sole focus of interest. Two pairs of Oystercatchers are displaying noisily, while a pair of Black-headed Gulls is showing interest in the 'tern table'. Best of all, Avocets are nesting on one of the rafts, and are well into incubation. Small numbers of birds typically begin to arrive at the Ponds in mid–late February. They first attempted to breed in 2007, but it is a hit-and-miss affair, with success by no means guaranteed. Our Ringed Plovers are a source of concern. They have been taking a beating from the local Crows (thankfully the Little Owl heard earlier in the year seems to have vanished). Last season we trialled putting protective cages over their nests, using a design developed at a sister colony across the Humber. The cages, box- or igloo-shaped, are made of a mesh with just the right size of hole to allow the plovers to pass through comfortably but keep most potential predators out. The plovers took to them with surprising ease: scurrying off the nest as we approached, they typically returned quickly, slipping through the mesh and settling down happily. Unfortunately, a local male Sparrowhawk cottoned on to what we were doing and took to perching on the cages and waiting the occupant out. This is not untypical of the battle against predators; it's often a case of two-steps-forward-one-step-back. Nor does it mean that because we had this particular difficulty, others would too. A more scientific and thorough trial later undertaken by Jake Taylor-Bruce of the Cumbria Wildlife Trust at Foulney worked very well, suggesting that our issue was with a particular Sparrowhawk, whose learned behaviour would not necessarily be replicated elsewhere by others. Given that these birds tend to live only for three or four years, there was a chance this one might not have survived the winter, so it seemed worth us trying again this season: we caged three nests and added the precaution of putting out five empty decoy cages to try to break the bird's automatic association of cage with food. It didn't work. It appeared that our hawk had survived the winter, hadn't forgotten about the cages and wasn't about to be fooled by the decoys. We quickly abandoned the trial.[9]

(The Foulney experiment was primarily aimed at whether it might be possible to cage Little Tern nests, but here its findings were more problematic. Unlike the plovers, the terns take off vertically from the nest, meaning cages have to be unroofed and therefore insecure. Stopping a Fox from jumping in, for example, would require an impractically tall cage – more a tube, in fact. Even then, it became apparent that a sitting tern, startled during the night by an approaching predator, runs a serious risk of flying into or even being caught in the sides of its cage. It's not impossible that a safer, more sophisticated model might one day be designed, but for the moment Little Tern nests remain uncaged.)

With the cages gone, the plovers again became more exposed to Crows, and losses continued. Evidence from the last few seasons suggests the Crows might well lose interest once the plover chicks hatch and become mobile and difficult to catch, but in the meantime the egg harvest continues. As with the Sparrowhawk, it seems that we are seeing learned behaviour – and few birds learn as well and as quickly as Crows – from two (collaborating?) individuals. Later in the season, if things run true to form, we might have as many as 20 young adults feeding on the edge of the colony but largely ignoring not only the plovers but other shore-nesters, including the terns. For now, all we can realistically do is scare off these intelligent and persistent predators as we best we can.

The thought inevitably occurs – and will occur again later when the terns settle down properly to breed – that shore-nesters seem do little to help themselves. By the second week of the month two plovers have nested outside our fences. The first found a home between the electric fence and the blue-rope perimeter

---

9 Experiments with Ringed Plover cages are not new: years ago they were tried at Gibraltar Point and Skidbrooke in Lincolnshire and worked well. Unfortunately, an item on them on BBC Radio 4's *Natural History Programme* gave rather too much away: sixteen clutches were soon taken by egg-collectors. 'Gib Point' continues to have form with cages: this very year, no fewer than *seven* species of raptor, including two rare Montagu's Harriers, have used them to perch on at different times, though not to hunt from. An Osprey employed one as its dining table.

*A Carrion Crow perches on one of our fenceposts, wholly untroubled
by the improvised measure we have taken to try to deter it…
Historically, the species is responsible for more losses of chicks than
any other, with the victims overwhelmingly being plovers.*

fence. We were able to move this incrementally the few feet into the colony
proper, to offer at least some protection from land-based if not avian predators.
The trick is to memorise or, better, photograph the nest in its physical context,
then edge it along bit-by-bit along with its surrounding 'markers' to replicate its
original surroundings in the new, safer location: that bit of brick just *there*, the
chalk pebble *there*, the plastic bottle cap there… As with the cages, the plovers
are tolerant of this kind of helpful interference, but on this occasion, it was to
no avail: the nest was predated by Crows.

The second nest was on the beach, outside the perimeter fence. This time we simply extended the fence, with its signs, around it, to advertise the presence of breeding birds and to provide protection from the careless feet of passing walkers (camouflage can be a mixed blessing) and their dogs – assuming they were on leads, which is by no means always the case. None of which accounted for, nor helped in any way with, the noisy presence of a group of quad-bikers who, despite being politely asked to desist, drove up and down the beach throughout the day and into the evening. Without local authority permission, it is illegal to drive motorised vehicles on public beaches. Repeat offenders can and do have their vehicles confiscated and crushed. James explained all this courteously and was just as courteously ignored. The nest, however, was still lost, though not to disturbance but to a predator, this time an unknown one. Of the four active nests from last week, one remains. Still, four or five pairs continue to hold territory and display inside the fences: bloody-minded persistence is as powerful a breeding strategy as any.

Human disturbance comes in many forms, sometimes unexpected and dramatic. The disposal of a Second World War bomb a few hundred metres down the beach this month certainly falls under the latter category, though not quite the former: it has happened here before, in 2011, and is not altogether surprising given that relics of the war still are still everywhere to be found in the area and on this coast generally. But these kinds of exceptional event are not our main concern. Sitting as it does between large caravan sites to both north (2.25km away) and south (1.3km), the colony is vulnerable to a wide variety of human disturbance, though the greatest intensity of beach usage is concentrated in the immediate locales of the sites.

Most wardens will tell you that their biggest issue is with dogs: specifically, dogs off-lead. Few dogs seem able to resist charging into a flock of shorebirds, if only – apparently – for the thrill of the chaos caused. And sometimes it goes beyond that, to deliberate and generally fatal attack. Either way, the damage to nesting colonies can be immense, up to and including outright failure. The problem became pressing during the pandemic, when dog ownership (and dog abandonment and theft) rocketed as the furloughed sought to get out into the countryside. Of course, as is often remarked, the problem is less with the dogs than with their owners. Contemporaneous surveys showed that very few people had more than the most basic understanding of the Countryside Code; one multi-agency exercise in the Lake District revealed that 87 percent of visitors were quite ignorant of it. In the North East, a group of (mainly) dog-walkers threatened to go to court to try to establish three new rights of way around Budle Bay as a result of being 'shut out' of a tern colony. It is possible for councils to impose Public Space Protection Orders (PSPOs), but few have the resources to provide the staff necessary to police protected areas. While the law may require dogs to be kept under 'close control', it does not specify that that means being on a lead, leaving the definition of 'close control' unhelpfully open to interpretation and negotiation. And 'negotiate' is all we can do. Most – not all – people are reasonable and, once they understand what we are trying to do, willing to put their unleashed dogs on the lead until they have passed the colony – not least because they don't want to see their dogs get a shock from the electric fence (and neither do the wardens!). In the case of the more recalcitrant, we can point out that disturbing the colony, with intent or through negligence, is an offence punishable by law, though even that fails to persuade some, and can provoke aggression. At that point we can do little more than withdraw. It's important to remember at these times that we are dealing with a small minority, and that our encounters with the public (not just dog-owners) generally are opportunities to gain understanding and support.

On top of the terns and plovers, we now have two active Oystercatcher nests and the Black-headed Gull has laid and started incubating on the 'tern table'; we wonder if this is the same individual that laid there last year – a bird that sat for close to 12 weeks on one infertile egg. I suspect I'm not alone in silently wishing her better luck this season. If she's successful, it will be the first record of the species breeding at the Ponds: recent attempts have all failed. Best of all (especially given their own failures in recent years), our Avocet pair has hatched at least two of four eggs.

And the Little Terns are becoming an ever-more visible and active presence. Thirty-one birds were seen on the 10th, but numbers were usually a little lower, and there were as few as eight at one point. It is the following week when numbers really take off, with between 40 and 70 typically present around the colony. This is the week in which arrivals have surged nationally. Inland sightings, almost always of single birds, are being reported daily from across the country, but also, unusually, two large (for Little Terns) groups have been seen, with twelve reported from Nottinghamshire on the 15th and nine from Lincolnshire on the 16th; it's tempting to speculate that these might be the birds which have helped boost our population.

Nest-scraping has also picked up. Three pairs began, rather desultorily, a week ago, but now that numbers are rising more birds are scraping away in earnest.

Little Terns in flight, with half a dozen Dunlins and a Little Gull.

Nest sites are often similar to those of previous years, but it's thought that – unlike other tern species – Little Terns do not necessarily stick to the same partners from year to year. Again, no-one's quite sure. What is true is that (again unlike other terns) nest sites are quite widely spaced: there are usually between two and five metres between them, depending on overall available space; the eggs are, logically enough, rather better camouflaged then too. It's notable that when the colony is 'squeezed' and populations become too densely packed, the level of aggression between the terns rises, with adults attacking their neighbours' chicks if they stray too near the wrong nest, whether those chicks are other Little Terns or other species. In the 1999 season a pair of Little Terns was watched killing a Ringed Plover chick that had wandered too close; they continued attacking the corpse for an hour. Little Terns themselves can be the victims too: recently, an Oystercatcher was seen to catch one by the wing and shake it violently before it escaped.[10] So proper spacing between nests is

---

10 Oystercatchers can be particularly aggressive, but the incident witnessed by former estate worker Jonnie Fisk a couple of seasons ago, of a bird attacking and killing (by drowning) a Redshank, was exceptional. According to Fisk, other waders present showed "a morbid interest" in the event.

crucial in helping minimise aggression between nesters; but it might also prove to have another possible, if unexpected, benefit later in the season, as we will see. And the whole question of where Little Terns nest in relation to one another and to any vegetation present is one that Ollie, our University of Leeds MSc student, based his dissertation on last year.

Mornings and evenings are usually the more active times for the terns; the middle of the day tends to be much quieter, giving wardens the chance to check up on other birds: our second Oystercatcher pair (outside the electric fence) has lost its eggs to a Fox (judging by the tracks around the nest), but a new nest has been found at the south end of the colony. Plovers continue to display, but no new nests have been located, and one adult has been taken by a Peregrine. On the other hand, Crows thankfully – and not untypically – seem to be losing interest now, and although both Sparrowhawk and Kestrel are seen regularly around the site neither has yet made any real attempt to hunt. The Avocets did hatch all four eggs, though the fourth chick, which emerged a day after its siblings, was quickly lost. The parents took the remaining three into the vegetated dunes at the south end for better protection – typical behaviour for this species. Happily, a second pair has been seen sitting nearby.

An Avocet brooding its chick. The species has only a short and patchy breeding history at the Ponds, but continues to maintain a toehold here.

I make an off-rota visit late one afternoon mid-month: it's a pleasant day, a warm sun tempered by a cool breeze. The colony is quiet. Plovers, Redshanks, half-a-dozen Avocets and as many Oystercatchers are loafing around. A few Sanderlings and Turnstones are busier, and a couple of Arctic Terns fly through. A Linnet, its bill stuffed with insects, drops into the dunes behind me.

There are just six Little Terns, huddled together on the shore of the Ponds.

Suddenly they rise in excitement as more birds appear: 10 more, then 20-odd, then nearly 50. They've come in from the sea, some bearing fish, yet they *seem* to have simply materialised out of the empty air. The whole character of the landscape is spontaneously animated, electrified, by their lunatic chasing and calling. They pass from entirely absent to overwhelmingly present in an instant, a burst of explosive, cacophonous energy, like children erupting into a playground at play-time. (Swallows can have the same effect, though their voices are sweeter than the squeaky-toy calls of the terns.[11]) They are infuriatingly difficult to count in flight, but at some invisible signal tension drains from the flock and they separate into smaller groups quietly to drop down and gather again on the shore and I can count them. Then they're suddenly up again, swirling high and noisily, only to relax and settle just as suddenly once more. This behaviour is known as 'dreading': it may be a response to predators or disturbance but as often seems impromptu and motiveless. The pattern repeats. A few drop inside the fence to mark their spots, but the majority continue to gather on the shore between 'dreads', perhaps engaging in some breeding process or hierarchical sorting invisible to my eye. They're still at it when I leave, prompted by my suddenly noticing a distant bank of threatening cloud roiling up.

In fact, the weather throughout the month has generally been dry and warm, the winds predominantly from the southwest, so migration has moved on

---

11 Terns, of course, are generically known as 'sea-swallows'. According to Christine Jackson, specific vernacular names for the Little Tern – unsurprisingly predominantly from East Anglia – include 'chit-perl', 'dip-ears', 'little darr' and 'shrimp-catcher', all of which are of Norfolk origin. Sussex chips in with 'little kip', 'scurrit' and 'sea-martin'. Ireland's 'fairy-bird' echoes the proper name of the Little Tern's close relative the Australian Fairy Tern; the name might well have been imported by human migrants, though I rather hope it was found independently, confirming the ethereal, almost magical quality of *Sternula* species in flight.

apace. On my next visit, a morning shift, there are fewer ducks on the Ponds but those that are there are now in their summer plumage, as are the Little Grebes. Swallows and martins hawk over the water, Pied and Yellow Wagtails pick their way round the edge, impossibly tiny next to the Little Egrets sporting their elegant summer headwear. Passage Wheatears on their way north have been moving since late March, but latecomers still appear out of nowhere on our fenceposts. On the walk in, the hedgerows hold Common and Lesser Whitethroats, while Skylarks and Meadow Pipits sing overhead. The colony now has an air of bustling activity, of colour and sound, that seemed impossibly remote barely six or seven weeks ago.

The first of our residential volunteers, Tom, has arrived: he'll be with us to the season's end. The electric fence was finally completed last week and is due to be switched on now that the terns have begun to settle. True to form, a problem has immediately occurred: one of the main strainer posts has snapped and had to be replaced. A second group of quad-bikers appeared on the beach, but proved more understanding of the need to stay well away. More worryingly, news has come from the south side of the Humber of suspicious activity in three protected sites: two men, claiming to be doing survey work for the RSPB, were found looking for Marsh Harrier, Bittern and Bearded Tit nests. It was quickly established that they were nothing to do with the RSPB, and the suspicion was that they were egg-collectors. A report, including descriptions and a car registration number, was passed to the police, and an alert put out to vulnerable nearby sites.

Egg-collecting has become much less of an issue generally since it became an imprisonable offence under the Countryside and Rights of Way Act 2000; only a pseudo-'professional' hard core remains active, and then mostly away from the public eye: the isolation of Scotland, for instance, where rare British species such as Capercaillie, Crested Tit and the endemic Scottish Crossbill have their strongholds, as well as the ever-glamorous raptors, is particularly attractive. What motivates these men (and they are pretty much exclusively men) is difficult for the outsider to comprehend. There are still limited opportunities to make money from the enterprise, at least if we include the lucrative American and Middle Eastern markets, but the real impulse seems highly individual and pathologically obsessive, and not even to do in the first instance with *possessing* the eggs; rather it is in the act of collecting itself that the kudos lies – the thrill of the 'treasure hunt'. This might explain why multiple eggs of the same species are often taken, with the same sites revisited repeatedly.

In recent years, and in contrast to the depredations of the more remote past (see Introduction), egg-collecting does not seem to have been a huge problem at the Ponds, certainly not since 2000. (Egg thieves were held partly responsible, with predation, for the near-total failure of the colony in 1989. The last recorded and confirmed case I can find was in 1993, when 'eggers' – betrayed by the footprints they left – took one complete clutch of tern eggs.) We are, for one thing, quite a public site, and certainly a very well-watched one. Inevitably there are suspicions from time to time, and individuals have been challenged; it is also said that confiscated egg collections in the Yorkshire area have contained Little Tern eggs almost certainly gathered from Spurn. There seems, however, to be a new trend developing: collecting eggs not for their own sake, but to incubate them and raise the young domestically, and then to breed from those young for the benefit of those who collect the birds themselves, rather than simply their eggs. The range of wild birds kept in captivity, and not just the more obviously glamorous ones, is astonishing. It is for this reason that we put ourselves on high alert.

The last week of the month saw the number of terns continuing to fluctuate quite wildly, with birds still spending much of the middle part of the day away. This is not unusual for late May, and there may be several reasons – all related to breeding; after all, that's what they're here for. As I've already suggested, it's thought that going elsewhere to display is one safeguard against attracting predators; birds also often need to spend time feeding up and getting into top physical condition after their long journey. That might be why, on my most recent shift, I saw females apparently refusing to accept the fish that many males were bringing in. I followed one bird as he came in off the sea, a good-sized sand eel dangling from his bill.

He quartered the colony for some time, apparently looking out for a receptive female. Eventually he landed and went into his courtship dance, circling his chosen target repeatedly and waving his prize. She however seemed to show little interest, walking away – though not far. And she sometimes started to approach and lunge at him. Was she irritated by his advances or offering herself, in which case was he now deciding on closer inspection she wasn't up to scratch? Either way he moved on to another bird and started over, but again with no result. Then a third, which seemed to approach him deliberately... only, fairly clearly this time, to turn him down. It's a complex business, Little Tern courtship. Both have to be ready at the same time. Our male finally admitted

Bringing in a Sand Eel.

defeat and went to stand on the Ponds' edge with three other males, each with a rapidly drying sand eel drooping from its bill: a picture of dejection.

Either or both sexes might have felt they were not quite ready to breed, not quite on top form. But the females at least might also be purposely delaying for other reasons: spring tides at the end of the month could wash away eggs laid too early, and experienced adults may well know this. There's some evidence from the 1990s that this has been a problem before at the Ponds, where the nests of what were thought to be first-time breeders were washed away while older birds held back.

This could become a particular issue down at the peninsula. As we've seen, Spurn Peninsula historically provided nesting-grounds for Little Terns, but these have been all but lost. Birds may still gather at traditional sites such as the Point itself as part of their early-season 'away days', and are sometimes seen apparently prospecting in small numbers for nest sites there and elsewhere. If, as seems likely, the southward slippage of the Ponds colony continues – and we are doing all we can to help the terns maintain at least a foothold there – then, with proper monitoring and protection, there seems no reason why they shouldn't recolonise the peninsula. The likeliest site now, however, is probably the Breach. Formerly known as the Narrows, this is the strip of land that connects the peninsula proper to the mainland just south of Kilnsea. In 2013 the North Sea broke through, and when it receded, access to the south, at least for all bar the most rugged and essential of vehicles (those belonging for example to the RNLI station at the Point) became impossible. Although for the great majority of the time it is a straightforward if slightly arduous walk or cycle ride, the possibility of a high tide, driven higher by those northerly winds, remains, and the owners of the area, the YWT, have not only put up warning signs but built a small hut on the south side to provide shelter for those unfortunate enough to be cut off. More positively, what the Breach (also known locally as the Washover) has provided is good habitat for breeding shorebirds. Ringed Plovers are already establishing themselves there, and at least one Little Tern has been seen scraping this season.

Contingency planning for the terns' possible relocation has already begun, and a key element of that is the erection of yet more electric fencing. (We finally switched our own on at the beginning of the week, having replaced the snapped strainer. So far, it's behaving.) A number of us gave a hand one weekend helping YWT wardens, interns and volunteers to start to put up a large enclosure on the Humber side of the Breach around the best-looking habitat, where the birds were showing interest. The need for it was immediately made evident when we noticed Fox tracks before we'd even sunk a post. Fortunately the Breach here is wide enough to accommodate both a colony and plenty of space to allow visitors – and there may be hundreds in midsummer, at peak breeding time – to pass by without undue disturbance to the birds. Or so we hope…

YWT interns Harry and Clare have joined the volunteer pool for the Ponds, too. James can now draw on half a dozen-or-so local volunteers for weekly or occasional support. The hard graft is done by the wardens and residential volunteers; we locals are there more to offer back-up, working shorter shifts, plugging gaps and concentrating less on precise monitoring and data-gathering than on public engagement (including the problematic kind) and simply keeping an eye out during quieter times. The colony is now covered throughout the daylight hours: activity is increasing daily – and not just because of the terns. The ever-beleaguered Ringed Plovers have now managed to get three nests going in the colony with one more on the adjacent beach, and we know there are several more further along the beach, outside our area of control. There are four active Oystercatcher nests at various stages of incubation, including one on the beach beyond our fences which is disturbed by every passer-by but persists nonetheless – at least for now. A second pair of Black-headed Gulls has begun nesting, and our Avocets are doing well, with the first pair's three chicks surviving and indeed prospering and the second pair on eggs. Human disturbance has been largely restricted to dogs off-lead. Predator-wise, Crow activity remains low, although a Kestrel, while posing no immediate threat, seems to be showing increasing interest. Around the colony's edge, Linnet and Reed Bunting nests have been found, with the former on the point of fledging and the latter freshly hatched. More Linnets and Reed Buntings are singing to claim their territories nearby, and have been joined by Sedge Warblers. Although these fall outside the immediate remit of the project, we'll continue to monitor them and submit our records to the annual Breeding Bird Survey run by the British Trust for Ornithology (BTO).[12]

Keeping on top of this increasing activity means more work for James and Rob. Weekly wardening rotas need making and, crucially, for accurate information to be shared quickly and efficiently across the team. This used to be done simply by written log, left in the Old Hut and filled in at the end of each shift, with the main points passed on verbally at hand-over. Now we use a dedicated WhatsApp group, which ensures everyone gets the same information at the same time. In the past this supplemented the written log, but now it has largely taken over from it. This came about largely because of the pandemic of 2020–21, where lockdowns forced us to rethink much of our practice. Initially we were afraid we would not be able to run the project at all, but we were fortunate in having four people available who shared accommodation in two pairs and who could therefore work together isolated in those pairs. With careful planning, and by adopting Defra and RSPB protocols regarding safe practices – not sharing equipment, maintaining distance when necessary, masks, disinfectant etc. – we

---

12 Historical records of breeding in the Ponds area are incomplete and unreliable, but still tell us something about how much the landscape and habitat have changed: Swallows, for instance, bred in a Second World War pill-box now lost to the sea, and Wood Pigeons in a plantation that no longer exists. But the data also hint at how many bird populations have fallen dramatically. Corn Buntings, for example, have declined horribly over the last half-century (mainly down to intensive farming practices), so while there is good habitat immediately around the Ponds, the last records are from 1999 and 2000.

RB 31/5, 1200–1700

13C, SE f2, cloud cover (cc) 7/8, steady rain, sometimes heavy, occasional thunder. Visibility poor. Easing, some sun by 1600.

F 5.8; 9.3 after rain

T 12.3[13]

LTs – none on arrival. 2 over 12.45, 3 landed w. Common Tern 1400, 1 over 14.45. 5 in-off 15.10, 3 with sand-eels: much chasing; 12 at 15.50, more chasing.

RPs – mating NW of Old Hut. Others on territory as per.

Oycs – lot of noisy territorial behaviour involving up to a dozen birds by shift end.

Avos – 4 adults S end. Chicks not seen.

BhG – both pairs sat all afternoon.

Predators – Crows in SW corner showed no interest. Ditto 4 herons. No raptors.

Disturbance – none

Other – 2 Sandwich Tern, 2 Turnstone on raft, 3 Dunlin, 1 Hare N end

were able to sustain a skeleton service. (As lockdown meant that fewer people were using the beach anyway, this proved largely sufficient.) A shared written log was clearly a potential source of cross-infection, so the WhatsApp group took over and, after some fine-tuning of data requirements and standardization of presentation, has remained our key means of sharing information.

Leaving aside the rain and thunder which have so far been mercifully light and irregular this season, this represents a pretty typical early-season

---

13 F = main electric fence, T = 'Trip' wire, measured in kV.

afternoon shift, with the possible exception of there being nothing to report under 'Disturbance'. The basic format is fixed and so easily maintained, cross-referenced and checked. It's important to get your shift report done quickly so as to maintain a clear overall chronology, necessary for monitoring how the colony is progressing day-by-day, for identifying and nipping in the bud specific problems and for noting particular behaviours. 'Things to keep an eye on.' It's also useful for raising and discussing more general issues without the need to convene meetings: most of our discussions earlier in the year about the use of Ringed Plover cages, for instance, took place here. As the season develops, shift reports will inevitably become fuller and more complicated, and emphases will change. Most notably, this year we will become a full part of a much wider survey of beach usage in the area, run by the Humber Nature Partnership (HNP). Having clear and detailed data on the frequency and nature of human disturbance to wildlife is crucial in arguing the case for greater protection, for funding applications and so on, so we are happy to be taking part. Much wardening work in any case involves looking as much outside the colony boundaries as looking inside.

And inside… frustratingly, it feels like the slow start to the season will never pick up. Having taken their time getting here, the Little Terns seem now to be also taking their time getting on with breeding. Some of the more southerly colonies are nearly three weeks ahead of us. There are already chicks running around at Chesil Beach in Dorset, and wardens at Winterton Dunes in Norfolk have counted over two hundred nests, with the first egg laid a week ago. The tern's incubation period is 18–22 days. For most of the recent history of our colony, the first eggs have been laid between 20th and 30th May (a week after spring tides) and have hatched between 10th and 20th June – subject, of course, to the vagaries of weather and predation. Chesil's chicks must have come from eggs laid early–mid May. It is not unusual, and is even expected, for us to be later starters than the more southerly colonies, but for the last two or three years we've been later still. Cold springs are the likeliest reason, but it's worth watching to see if a trend is developing – and why.

JUNE

MT 7/6/22, 0600–1400…

LTs – first day it's actually felt like a breeding colony. Numbers building from 30 to max of 74. Lots going off inside fence with up to 30 birds dropping in. Lots of displaying and scraping. None sitting for long in any spot but reckon, all being well, there should be birds obviously sitting in a few days…

For almost all of the first week of June things have continued much as they have for the last two or three weeks: fluctuating numbers of terns, doing all the right things in the mornings but then disappearing for much of the rest of the day and mostly not being seen to return in the evenings. The Ringed Plovers have lost two nests but gained two more, so we remain at four active nests… there's another Oystercatcher nest, making five… Avocets and Black-headed Gulls continue as before… the Sparrowhawk (still using our decoy cages for perching) and a male Kestrel are lurking without doing real harm, but a Peregrine was seen to take a Turnstone… Crows might have been responsible for the two failed plover nests (we are not sure), but generally don't seem very interested. Dogs off-lead continue to be our main worry, and there was a troubling, albeit second- or third-hand, report of one almost catching an Avocet chick. Worse still, a seal pup was found on the beach opposite the north end of the colony with bloody puncture wounds on its head and flippers and the tracks of a dog nearby. There was no sign of the animal or its owner. Though apparently malnourished, the pup managed to escape to the sea when Marine Life Rescue attempted to take it into care.

So MT's shift report was particularly welcome. Mick started on the first of the month when Rob also became full-time with us, so now we are at full strength warden-wise. He's a local man, though originating in the South Yorkshire area and one of that new generation of working-class birdwatchers (as we were

Monitoring the colony from the New Bank Viewpoint.

known then) who were largely responsible for the wider popularisation of the hobby in the 1970s; several subsequently moved to the Spurn area to live. Mick is greatly respected as a birder far beyond Spurn, and our most experienced warden, having already done the job for seven years. His responsibilities will involve mainly night shifts later in the season (last season his young grandchildren asked him if he was 'nocturnal'), but they won't start for a little while yet.

I haven't got a shift this week, but I'm prompted by Mick's optimism to walk a circuit of the colony anyway. For a change I approach from the south, the Kilnsea end, along Beacon Lane. It's a lovely, warm, late Spring day, complete with calling Cuckoo. The hedgerows flicker with butterflies: Red Admirals and Small Tortoiseshells mainly, with the occasional Common Blue and Painted Lady. I pass the local Kestrel's nest (of which more later); it's quiet at the moment. Whitethroats sing from the hedge tops, and I stop to watch a particularly vocal male. He might well have flown as far as the Little Terns, and possibly from an area right next to their winter quarters off West Africa. I've always taken a particular pleasure in Whitethroats, not simply because of their inherent charm but because it is a species I discovered late, when I returned to birding at the start of the 1990s after a 20-year gap. I should have been familiar with it from my childhood in the Cheshire countryside, but I wasn't – there were too few then to snag the attention of an impatient young boy: drought in their wintering grounds in the Sahel severely diminished their numbers in the late 1960s. A reminder, if one were needed, that the lives of our migrant birds are subject to pressures far beyond our immediate control. Perhaps surprisingly, that drought might also have affected the Little Terns: without the rain to wash nutrients into the rivers and thence into the estuaries, the sea's productivity falls, impacting on fish and those that feed on them.

My stroll north continues. The hedges peter out as I reach the end of the lane, a short stretch of low, scrubby dune, and I jump down on to the beach. I flush a large Hare; she "startles like a sudden thought", in John Clare's lovely phrase, and scampers away. She's the mother, perhaps, of the two young who were seen suckling inside the colony fence last week. A large pale mound on the edge of the dunes resolves itself into a shirtless middle-aged man as I approach. He's lying on his back, completely still and for a ghastly moment I think he might be dead. But then he stirs, idly scratches his stomach and settles down again. Sunbathing, I reflect, is one of the few forms of potential human disturbance with which we seldom have to contend. I'm relieved.[14]

It's not easy to stop wardening ('Raptor!' 'DOG!!') and of course if an issue arises you can't ignore it, but I still want – need, perhaps – to experience the colony in a more relaxed way; not to have to watch for predators or dogs off-lead, check the fence, count the birds. Rob and Harry are on duty anyway (we double up sometimes), so I can slow down, open up, take it in. Hence my untypical approach: I'd normally walk in from the north, along the beach, or from the west, past the Wetlands and along New Bank. Coming up to the colony from a different direction might help give me the shift of perspective I want.

I'm still well to the south of the site, though. So I set off, dropping first down to the tideline. It's a calm day, but the southwesterly wind, gentle as it is, is still pushing the incoming tide into small breakers. A huge bull Grey Seal, considerably bigger and heavier than me (and I'm not little), breaks surface a few metres out and eyes me for a moment before sliding under again. Seals

---

14 Although it is true that on a couple of occasions in the past wardens have disturbed reconnoitring nudists in the dunes. Naturalist meets naturist. And our colleague across the Humber, Liam Andrews, once had a nasty turn finding a body on the beach. Thankfully, it turned out to be a fully clothed mannequin, a practice dummy lost (a little worryingly, it must be said) by Air-Sea Rescue.

are renowned for their curiosity, but this one clearly found me entirely uninteresting. There's something reassuring about my irrelevance to him.

The lulling rhythm of the lapping waves and the calls of the occasional gull serve only to draw attention to how still and quiet it is generally.

The beach seems deserted, but then I start to notice Ringed Plovers scuttling away: one or two at first, then a flock of a dozen or so (I consciously resist a proper count) explodes from invisibility, peeping alarm as they fly low up the beach and resettle. I begin to see more: a few Dunlin, seemingly more relaxed than the plovers but still quietly picking their way steadily out of my path, their nonchalance feigned; a lone, brilliantly white Sanderling, skittering along the water's edge.

I'm level with the southern end of the colony now, at the point where the fence emerges from the dunes and right-angles to run north, but I stay as far away as I can, following the ebb and flow of the tide: like the Sanderling, playing at risking wet feet. And on cue they come: seven or eight Little Terns, riding high and noisy in from the sea, somehow lighter than the air they are capering in. At least two males, I think, their bills glinting silver in the sun with the sand eels they are carrying to impress the females (the bigger the better, it seems). I've already missed most of the aerial courtship that will have been taking place, but I've seen it often enough before. The male rapidly spirals upwards, a female following his flight. At the top of the ascent he drops into a downward glide and is overtaken by the female in a manoeuvre known as the pass, the closest the birds come to each other. She's checking out his fitness and suitability as a mate. Both birds weave down to the ground, repeatedly crossing each other's paths. It's a dance, as formal, intricate and expansive in its way as an eighteenth-century *minuet* and, like a human dance, within the formality is passed subtler information not obvious to the outside observer. With the terns this is believed to involve, surprisingly perhaps, flashing colour. A bird in breeding condition, freshly moulted, reflects colour at the ultraviolet end of the spectrum, invisible to the human eye (it's becoming apparent that a number of species make use of UV in this and other ways – your male garden Blue Tit for one). I have often wondered whether that stands simply as a general indication of breeding fitness, or if more complex, nuanced signalling is taking place? Something we might call… flirting?

And this is only the high courtship flight; there's a low-level one too, with its own rules and rituals.

Mid-courtship

But I'm concentrating on the two that have dropped to the beach to continue the ritual on the ground. It's full-on now, with none of the uncertainty of earlier encounters, and pretty much textbook: both birds assume a head-up pose, wings half-open. The male bends down and starts to circle her. She starts begging like a chick, her body shivering. Sometimes he'll hand the fish over now, but not this time; he approaches her from behind as her shivering increases, then jumps on her back, waggling his fish above her. They continue this for a little while, then he lowers his tail, and they make sexual contact just as she takes the fish from him. It's done. According to some authorities, sitting females waiting for their partners to bring fish will sometimes go through this entire routine with another male who fancies his chances, only to shrug him off in the split second between taking his fish and sexual contact. On the other hand, we have an historical report of a male removing an egg from the nest of a sitting female with which he wished to mate. All's fair... No such shenanigans for this couple: now they are a proper pair. I can breathe again. I'm barely 30 metres away. I've frozen to avoid interrupting them, though to be honest I'm not sure they've even noticed me. They were certainly oblivious to an inquisitive plover that briefly approached them.

I walk on to duck under the perimeter fence and enter the colony at its northern end, near the Old Hut. A day earlier and I would have had to have taken a longer way round to avoid disturbing a Ringed Plover that was sitting nearby; but the nest has been predated overnight and it's safe to proceed. I'm heading towards Rob and Harry who are at the Viewpoint on the far side of the Ponds, the better to observe what's going on across the whole colony. A female Reed Bunting flits across my path through the dunes, her chicks were ringed a couple of days ago. The Skylarks evidently have young too: one of the adults zips over with a bill full of food. But I'm not really paying attention now. I'm pondering the possible implications of what I've just witnessed.

Just because birds copulate on the beach doesn't mean they will nest there.

But what if they do?

I know that they have in the past, though in smaller numbers than within the fenced area. I am also aware that their track record on the beach is not good. The habitat, of course, is great: there is much less vegetation to avoid and plenty of fine substrate to scrape a nest in. But unprotected they are so much more vulnerable to predation and disturbance, and there's a greater risk of them losing their nests to high tides or even wind-blown sand. Are we in a position to extend our fencing and signage around beach-nesting birds? Will we be able to monitor the increased area well enough? And, crucially, how will passers-by, especially locals, view us taking part of the beach if we have to?

Most people are supportive of and interested in the colony and our efforts to protect it, but not all. I'm not completely unsympathetic to those who find our presence irksome: particularly when the pandemic was at its height, it must have been doubly irritating finally to escape the severe limitations of lockdown and get out into the freedom of the wild only to run into yet more 'dos' and 'don'ts' – and that was without fences and signs actually on the beach and well-meaning people with binoculars asking you, however politely, to retreat.

Less forgivable, though, are the birders and photographers who get too close and are seemingly oblivious to the alarm calls of the birds they have flushed and which are now flying over their heads. They should know better. Taken as a whole, the range of human disturbance is considerable: as well as those I've already mentioned, on the beach there may be dirt-bikers, fossil hunters, fishermen (beach-casters), ramblers and hikers (sometimes in groups), joggers,

Too close. Despite repeated requests to view the colony only from New Bank, some still approach from the beach side and disturb the birds.

cyclists, and detectorists; on the sea, canoeists, sail-boarders and (one year) radio-controlled boat enthusiasts; above, occasional low-flying light aircraft (and occasionally higher-flying but rather louder RAF fighters), microlights, hang-gliders, kites and drones. Thankfully, in most cases, even when the beach is crowded to the north and south, our bit in the middle is much quieter, though in 2020 a sunny post-lockdown weekend pushed big numbers of holiday-makers and day-trippers on to our part of the beach when an accident blocked the road to Spurn. An incoming tide exacerbated the situation, edged people closer together and closer to us and gave our volunteer warden Luke a particularly busy shift.

Of course, most of this disturbance is entirely innocent, and the great majority of people respond positively and even apologetically to our soft-touch policing. It's important we share our enthusiasm. We are always polite, and try to explain what we are doing and why: that these are Schedule 1 birds, threatened, and protected by law. If tension escalates, we back off and may radio for help, but it is more usual for folk to show real interest in the birds and the project and sometimes even walk away as converts to the cause. Incidents of malicious interference or real hostility are thankfully rare, and always reported to the police. Casual vandalism of the Old Hut is not uncommon outside the breeding season, although some years ago there was a more serious incident, combined with abuse and physical threat, during the summer (see Introduction). Something very similar happened at the Teesside colony last year. In 2018 a dog-walking family was seen to try, rather unwisely, to unclip our electric fence, presumably as an act of protest. Some incidents remain plain odd: earlier this very month a shirtless man walking down the beach decided to vault the electric fence for no reason that was discernible. When asked to leave he did so immediately and entirely happily and chatted away in a friendly way without offering any explanation. We're still none the wiser.

In the end the birds, on whom the distinction between innocent and deliberate interference is naturally entirely lost, must come first. We have to work with the fact that not everyone sees it like that, that not everyone is open to persuasion, and that there are limits to how far we can offer protection. People are entitled to use the beach, barring illegality, as they will. There are balances to be struck and we have to accept that. And so, in a sense, do the birds themselves: if we can mitigate disturbance enough, then their tolerance of it may grow to a degree. Maybe it already has.

I still hope they don't nest on the beach, however.

I'm walking quicker now, irritated with myself for letting 'wardenish' thoughts take me over. I shake them off and stroll on, making a deliberate effort to slow my pace. I reach the Viewpoint and the habitual banter that characterises the team's communications immediately kicks in: "Come to check up on us?" "Yup. Notes will be taken." (The exchange finds its way into Rob's shift report later. I'm under 'Disturbance'.) I pass on what I've seen on the beach, and Rob confirms that the pace is picking up in the colony. Almost 50 birds have hung around all afternoon, with at least ten pairs scraping and one pair seen copulating twice.

Most of the action is taking place in the southern half. There are also more Oystercatcher and Ringed Plover nests too (the latter still working hard to keep abreast or preferably get ahead of continuing losses) and there are now three pairs of Avocets (the first still hanging on to their three young and the other two pairs sitting) out of a splendid total of 19 birds present. Altogether there's a real buzz on and I pause to enjoy it as I continue back south along the Flood Bank. We're still at nothing like full throttle, but the grey cold of our winter visits is now far beyond imagining.

Back at home I check the records: the terns have often tried to nest on the beach but have seldom made a success of it. In fact, in one year at least (2004), the

An Oystercatcher about to settle on her nest.

majority of birds were reported as showing more interest in the beach than the usual colony, though in the end only eight pairs tried to nest there, with 23 pairs inside the fence. This is a repeating pattern. Over the next three to four days the terns, encouraged by the greater expanse of beach exposed by particularly low tides, gather in numbers to create what James calls 'a constant spectacle' of displaying, kicking sand, scraping, passing fish and mating: a combination of competition between unattached birds, pair-bonding and general infectious excitement. Almost a hundred individuals are now present, with 15 pairs in the colony – two of which are sitting though not yet, we think, incubating. As yet nothing has taken up residence on the beach.

But then…

By the end of the third week of the month we have a total of 15 nests, ten confirmed as incubating though not all yet with full clutches… and most of them are on the beach.

Nest-numbering is one of the innovations brought in this season by James and Rob. It may seem an obvious thing to do, but various factors – the nature of the terrain, sightlines, the difficulty of seeing the whole colony at one go – make it more complicated than you might think. We have in the past considered

*An adult prospects for a nest site. She eventually
settled here and produced two chicks.*

TW, 18/6/22, 1700–2200

Wind S f2, cc 6/8, 15–13C, rained for the majority of the shift

F: 7.7

T: 11.8 taken at 21.30

Little Terns: 7 birds sitting on the beach, 6 within the electric fence, 2 on
the edge of the Ponds. Was going to photograph each nest so people could
find them easier but the rain put a stop to that. Max count of 83 at 20.45.

Ringed Plovers: RP7, RP14 and RP15 sitting

Oystercatchers: OC1 with 1 chick and OC3 with 2. OC4, OC5 and OC6
sitting.

Avocets: AV1 3 juvs, AV2 3 chicks, AV 3 sitting.

BhG: BhG1 still 1 chick, BhG2 sitting

Predators: Kestrel S at 21.20 mobbed by Oycs

Disturbance: no people all shift

Other: 5 Hares including 1 that swam from the Flood Bank across to the
shore by the rafts twice!?

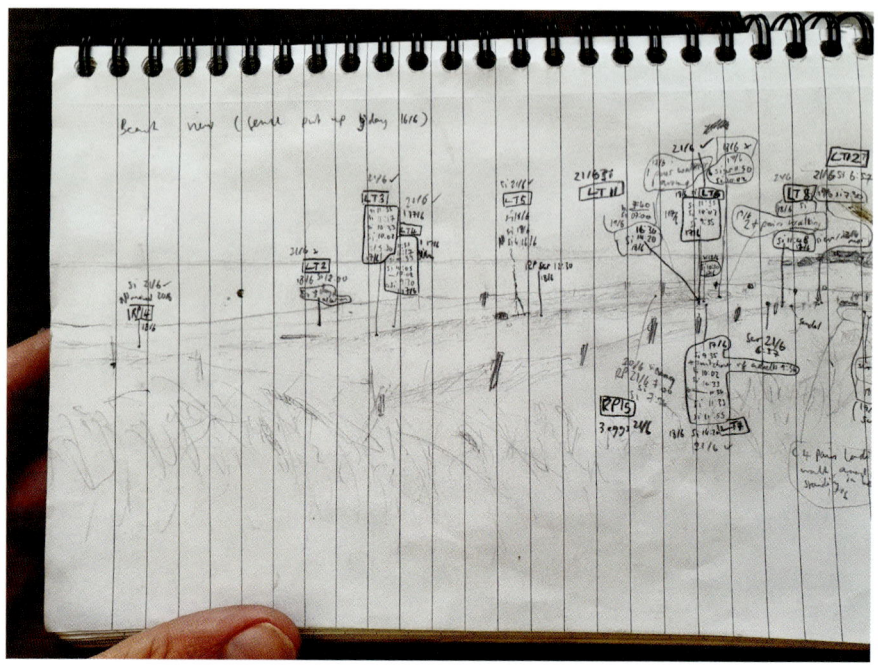

James's sketch map of numbered nests in late May, from the Viewpoint. The microscopic annotations (clearer in the original) show dates of laying, numbers of eggs, hatching, failures (and reason for failures if known), re-lay (if any) and so on.

other strategies to help us identify and share precise nest locations, including numbering electric fenceposts to narrow the area to check or even placing coloured stones nearby nests, but the former doesn't help that much while the latter might well aid eggers or even predators – especially the intelligent corvids. James and Rob's solution uses labelled sketches and photographs to show LT1, LT2, LT3 and so on, extending to RPs, OCs, AVs and BhGs.

This is a lot less clumsy than 'the tractor tyre Oystercatcher' or 'the tern by the south dune next to the tatty blue rope and pointy orange brick' (though sometimes we still have to resort to a bit of that).

It is quickly accepted that we need to extend our blue-rope perimeter fence, with its attendant signs, around the beach terns. We will add new signs, asking passers-by to keep to the tideline – that is, always as far away as the sea will allow. We're still not sure that the beach nests will survive the next high tides.

If they don't, the birds may re-lay in the same spot or nearby (usually after nine or ten days), move into the fenced area or down to the Spurn Washover, or even relocate to another colony entirely. We have persuasive, if sometimes circumstantial, evidence that this happens, as indeed the textbooks say it does: one year (1997), 30 birds deserted our colony under pressure from Crows and the colony on Teesside counted an extra 30 in theirs the following day. And in 2013, five Teesside adults which had suffered egg-losses arrived late in the season at the Ponds, with re-laying in mind. There's similar evidence of birds moving between the Ponds and the North Lincolnshire sites. (We know this because the birds in question had been ringed, of which more later.) If the beach-nesters do survive the tides, then we'll need to extend our electric fence in some way, and face what might be a few uncomfortable conversations with other beach users. This is on top of continuing problems with the existing fence: another strainer post has snapped, and a large mammal, probably a Hare but possibly a Fox, has apparently jumped through the fence nearby, displacing wires to leave an inviting gap. James and Rob have to attend to both quickly. The limits of the fence have also been exposed. The unexpectedly aquatic Hare is unlikely to have done much damage – certainly less than the Otter which Tom initially feared as he had spotted a distant head moving purposefully towards the colony might have done. (It's an unusual phenomenon, but not unique in the Ponds' history.) But the dead Hedgehog later found inside the colony might well have had a more negative impact: small enough to get under the bottom wire, it now seems likely to have been responsible for some at least of the Ringed Plover losses we had put down to Crows. Our colleague Mike Pilsworth of the RSPB has come up with a solution for the beach, however: we can borrow (from the Blacktoft Sands reserve near Goole) a sheep-mesh type electric fence, which will be quick to put up and if necessary take down, and should be manageable and effective. We plan to install it next week.

We've been joined by our second long-term residential volunteer, Murphy. And a good job too: not only are there the beach-nesters to deal with, but all the while the pace of life in and around the colony is increasing. Our first Avocet chicks, all three of them, have fledged. It's a major success for their parents and for us – this is the first successful breeding for several years. The second pair has hatched three chicks, though one died within three or four days; two more pairs seem to be toying with the idea of nesting without quite settling to it. We have six active Oystercatcher pairs now. The first to lay lost one of its two chicks; the second and third pairs each has two chicks, while the rest are at various stages of incubation. The first Black-headed Gull chick is coming up to two weeks old, resides permanently at the edge of the Ponds under the protection of at least one parent, and has been safely ringed. The second gull, however, has had some kind of mishap: eggshells could be seen by the water's edge, but there was no sign of any chicks. Predation is the obvious reason, but we simply don't know. James suspects the eggs might have been 'bad' and that the young died on hatching.

Other breeders appear to be having more luck. Reed Buntings seem particularly busy, with nests suspected behind the New Hut and in hedgerows near the colony. Skylarks, Linnets, Meadow Pipits and to some extent Sedge Warblers might also be found, although none is as reliable as the Reed Bunting and the success of all has been very variable over recent years.

A family of Shelducks has appeared on the beach and made its way on to the Ponds; the parents will do well to get all their 12 chicks to maturity. Wader numbers are growing, as non-breeders begin to return from the Arctic: Dunlin, mainly, but with the odd Curlew, Bar-tailed Godwit, Grey Plover and a rarer Curlew Sandpiper. Swift passage is starting, too. (This is pretty much an annual and rather mysterious phenomenon at Spurn: big movements of birds go south, usually from the end of June and onwards into July. Whether they are failed breeders leaving early, or birds following big cyclonic weather systems – in other words, flying in large circles – is a source of much speculation.) A flock of over 400 Starlings is feeding regularly in and around the colony. Rare visitors have included Black Tern and Honey Buzzard, as well as a Lesser Emperor Dragonfly and a couple of Hummingbird Hawkmoths.

As the level of busyness rises, so does predator interest, though actual losses remain small. Leaving aside the suspect Hedgehog, the only land-based threat we are sure is present is Fox, with tracks being found frequently – outside the fence. A Peregrine (again!) was the only raptor seen to succeed in taking prey, though we could not identify the unfortunate victim; up to three Kestrels are hanging around, but the Sparrowhawk has become a more occasional visitor. A Hobby flew through without showing obvious interest in either the birds or the dragonflies. As the so-far unrealised threat grows, so do the colony's defence mechanisms, with the terns, Oystercatchers, Black-headed Gulls and Avocets mobbing incessantly, though we feel a little concern that the Kestrels simply ignore all but the Oystercatchers.

Things settle down a little in the last week of the month. The new beach fence is installed and after a few false starts seems to be working reliably. Only one group of passers-by has taken real exception to its intrusiveness, but after chatting to James they proved more understanding. At busy times we've taken to doubling-up on shifts, stationing wardens at both ends of the colony. The birds themselves alarm when people pass, but are reassuringly quick to settle again. Combined with those inside the colony proper, there are now 28 pairs incubating, with the majority (17) inside. However, as feared, on the beach side we have lost one Little Tern nest and one Ringed Plover's to the high tide. Our attempt to relocate the tern's nest nearly succeeded: the bird returned to sit on several occasions but never seemed entirely happy and finally deserted. We had been unable to recreate the original well enough. The ever-stalwart plover did accept a moved nest, as did another tern, where we were able better to note the context of the nest and recreate it before the tide reached it. We have now found 18 plover nests since the start of the season, though all but five have been lost or become inactive. Three broods of Oystercatchers are present, with one having appeared out of nowhere with its single chick. Unhappily, all the Avocets bar the first pair, whose chicks continue to thrive, have been predated or otherwise lost in the last few days, and all the other adults have left the Ponds area.

That mystery Oystercatcher is a reminder that nesters can always be missed, however thorough and careful the counting is. The birds, after all, do not want to be seen, and from the camouflage of their eggs and young to the way they approach their nests their behaviour is driven by the need to remain unobtrusive. Even from our Viewpoint some sitting birds may simply be invisible to us. A bird repeatedly landing in a particular place probably means there is a nest there, but only probably – we might only be sure once we see feeding of young begin. All we can do is to record and share what we are able to see as fully and accurately as possible. James in particular is an absolute demon for data-gathering.

The numbering system doesn't, of course, aid us in actually finding nests, but it does help formalise and make consistent our shared mental map of the colony topography and, crucially, create an accurate history of the breeding season – what's sitting, incubating, hatching, feeding, fledging, deserted, been predated and so on, and *when*.

To count nests (more or less) accurately, and to check on their productivity, we need to go into the colony. This is something we do as rarely as possible (the other instance is when we ring birds) to minimise disturbance. We make only one 'pass' – if we realise we've missed something, tough – and use only small teams. I find it quite nerve-wracking. There is no threat to us: Arctic Terns are famous for relentlessly dive-bombing intruders and even drawing blood, but it seems to be no more than an occasional and ineffective tactic for the Little Terns. I've been 'assaulted' only once in five years, and only Rob and Murphy, I think, have witnessed it this year. The real cause of anxiety is the fear of standing on the superbly disguised eggs or young. It's not happened in my experience, but the grinding of fine pebbles beneath your boots can sound horribly like eggs being crushed. So we examine where we put our feet very closely, and move very slowly, spaced out in a line, and when we find a nest stop and raise an arm to bring over whomever has the ringing gear and the GPS locator.

We've been experimenting with GPS location for a couple of seasons: it gives us the opportunity to push our recording practice further, not only by pinpointing individual nests and their proximity to each other but also by getting a true overall picture of the size and shape of the whole colony and in particular how

Little Tern eggs in their scrape.

its historical drift to the south is developing. This is crucial for planning our future management of the site. It's clear that if we are to preserve it we will have to engage in some serious habitat management, and sooner rather than later.

That's for the end of season debrief and winter discussion, however. For the moment it's egg productivity with which we are mainly concerned. Depending on weather, food supply and the experience of the adults, Little Terns typically produce one to three eggs, laid one or two days apart and hatching within 24 hours of each other (the third sometimes a little later). At each stage it's a slightly shorter process than for their larger cousins, perhaps to minimise the time they are vulnerable to predators. Here are some of Rob's figures for the walk-through on 30th June (the first numbers are simply those of the readings):

149 – RP – 3 eggs

150 – LT – 2 eggs

151 – LT – 2 eggs

152 – OC3 (probably, based on location) – 2 eggs (likely both addled)

153 – LT – 2 eggs

154 – invalid reading

155 – OC7 – 3 eggs

156 – LT – 2 eggs

157 – LT – 2 eggs

158 – LT – 2 eggs

159 – LT – 2 eggs (1 tiny egg – MH photographed)

160 – RP17 – 4 eggs

161 – RP – 4 eggs

As you can see, it is by no means always possible to match nests found to those identified and numbered from the Viewpoint and the Huts. That's a task for later, and even then will probably prove impossible to complete definitively. Rob already suspects at least six tern nests were missed. Altogether,

25 nests were found of three species: 16 Little Terns, seven Ringed Plovers and two Oystercatchers (only one of which seems active). In terms of eggs, there were 30 tern eggs (at an average of 1.9 per nest), 23 plover eggs (3.3) and five Oystercatcher. It was disappointing that no terns had produced three eggs (this might be down to the late start or to the birds being in suboptimal breeding condition due to food shortages here or in their wintering grounds) though it's worth noting that the one tern which nested on the Washover did manage three eggs before the nest was predated; still, we'd very happily settle for a productivity of 1.9 at season's end.

As a half-term report, then, it's a bit of a mixed message. The Little Tern results are pretty much as expected, and in line with the colony performance over the last few years. Ringed Plovers, on the other hand, are so far having as bad a year as anyone can remember. And Rob makes one other observation in his notes: we found a dead Little Tern in the colony. It's probably the one 'rescued' by Tom a few days ago, when he heroically chased the length of the colony after a Sparrowhawk and eventually distracted it enough to drop the tern it had taken. The bird had seemed alright at the time, but might have sustained injuries which caught up with it later. We can't be sure.

JULY

To be honest, we've been holding our collective breath for a while.

Bird 'flu (Avian Influenza) is far from new to the world, but its various strains
have become an increasing problem since the early 2000s, with hundreds of
millions of domesticated birds dying or being culled between 2005 and 2021.
The numbers of wild birds affected are unknown. But with wild bird populations already declining alarmingly under the pressure of habitat loss, pollution
and, in the case of seabirds, overfishing, any susceptibility to so potent a disease
is very bad news indeed. In the UK, last winter's devastation of the wintering
population of Svalbard Barnacle Geese around the Solway Firth was particularly
shocking.

Bird 'flu is not new to the Spurn area either: it's been found in local wildfowl
on several occasions in recent years, though its exact extent remains unclear;
there has certainly been nothing remotely near the scale of the Barnacle Goose
outbreak. The virus is believed to have originated in the Far East in factory-
farmed poultry and to have been spread west at least in part by migrating wild
birds, perhaps in relay (the disease takes hold so quickly it seems unlikely an
infected bird could fly very far). At Spurn it is most likely to have arrived in
wintering ducks, especially Wigeon. The Obs was briefly involved in a putative
government scheme to catch, ring and test birds to understand better how they
move internationally – and where interventionist containment methods might
be tried on an international scale. But funding evaporated before the project
could get off the ground.

So far, bird 'flu has been a winter phenomenon: the virus seemed not to thrive
outside the body in warmer temperatures. But that appears to be changing.
What we are seeing now is a fast-spreading and particularly virulent summer
strain (H5N1),[15] which seems to be attacking seabirds, and seabirds that nest

---

15 The particular strain has been traced to southern China, and is believed to have
originated in a goose farm in Guangdong province in 1996.

colonially in particular. In UK terms, this seems at the moment to mean primarily Great Skuas, Gannets and Sandwich Terns.

This is why Tom's discovery of a dead Little Tern was so worrying: it opened the door to some troubling possibilities.

So far, evidence of the presence of the virus in the area this season has been inconclusive: a couple of dead Kittiwakes have been found on the beach, and a sick or injured Dunlin has been reported now for a couple of days. In any other year, neither incident would have excited much concern: birds catch diseases, get damaged and die all the time, and as Obs Head Warden Paul Collins points out, it is too easy to assume that every distressed bird is a 'flu victim.

Our dead Little Tern showed no obvious sign of the disease.

Just as troublingly, however, has been the video footage caught by a local birder of a Crow attacking and killing a visiting Sandwich Tern. A healthy Sandwich Tern seemed at first an unlikely target for a Crow: they are our largest tern and are agile birds, armed with a long and pointed bill. Could its vulnerability suggest debilitating illness? On the other hand, Crows are capable killers of other birds, given the opportunity; only recently one was seen taking a Snipe (admittedly a much smaller bird) nearby. Reports from Scotland, however, reassuringly suggest they have been known to take perfectly healthy Sandwich Terns when the chance presents itself.

But if there is a threat to our Little Terns, then it seems likeliest to come via transmission from their larger cousins. Sandwich Terns are the first of the terns to arrive back in the UK to breed, usually in mid-April, and they get on with the business quickly. The strikingly marked juveniles leave the nest site within days of fledging to disperse widely around our coasts, accompanying their moulting parents. With colonies to the north and south of us, their annual gatherings at the Ponds are a regular feature of our Little Tern summers from July onwards, with sometimes hundreds of birds being present. At their nesting sites, they associate closely with Black-headed Gulls and these too arrive at the same time in their hundreds: together they make a real spectacle. These birds could have come from anywhere, but our nearest colony is at Scolt Head in Norfolk, only some 80km away as the tern flies. The news from there is not good. Its large Sandwich Tern colony is being devasted: in a period of five days in mid-June, 56 adult Sandwich Terns were found dead, and half the chicks were dead or dying. (Twenty-one Black-headed Gull fatalities were also counted.) It seems inevitable that things can only get worse, probably much worse: the colony is thousands strong. And yet the neighbouring, much smaller, Little Tern colony is – so far – unaffected, with around 90 nests (and more pairs to settle) and many three-egg clutches. To the north, our nearest neighbour is Seaton Carew on Teesside (140km away), and its colony seems similarly healthy, with 32 Little Tern nests already hatching. Although Sandwich Terns do not breed there they are regularly to be seen, as are Common and Arctic Terns – all potential transmitters of the disease.

This is a pattern that seems to be established across the UK: losses to Great Skua, Gannet and Sandwich Tern colonies in particular are horrific, with other species – auks, gulls – also very vulnerable. What makes this outbreak particularly invidious is that seabirds are long-lived and slow-breeding: damage to populations takes years to repair. But Little Terns are so far untouched and even, in some colonies, such as Gronant, the largest breeding colony in Wales and home to 10 percent of the entire UK population, doing better than ever: it currently has over 200 nests, a record for the site. (The big Irish colony at Kilcoole is also thriving, with over 250 active nests and 430 chicks.) Why Little Terns should be unaffected is unknown. It may be that the greater distance between nests helps: Sandwich Terns breed pretty much bill-to-bill, and their colony hygiene is frankly appalling. Then again, Great Skuas, the worst affected of all (their entire UK breeding population, limited to the Northern Isles, may well be at risk), might be colonial nesters but in a looser sense: they keep their nests tens of metres away from each other. And although Little and Sandwich Tern colonies do not intermingle, they may nonetheless be very adjacent to each other, as at Scolt Head.

All this assumes that transmission of the virus is through direct contact. This is likely the case, but it seems to not be absolutely certain and it does not neces-sarily preclude other forms of transmission – through water, for example. Other aspects of the Little Tern's behaviour might play in, too: it is fast-in, fast-out as a migrant breeder, seldom lingering here after finishing its business. It also, once settled, stays local, typically fishing within no more than five kilometres of the colony (I live on the coast seven kilometres to the north; I'm lucky if I see one of our birds there per year). And they typically feed on smaller prey items than their larger cousins, so need not compete to share fishing grounds? Therefore, contact with other species, though clearly inevitable and regular, might be less frequent? Perhaps putting Little Terns at lower risk?

It's all guesswork at the moment. An anxious time.

There is one other thought, which we hardly dare speak out loud: might the Little Terns, alone of all their kind, be… immune? It certainly seems the case that their very close congener the North American Least Tern is also escaping the ravages of the disease. Might all *Sternula* species be immune, at least to this strain of the disease? Will they remain so if, when, it mutates?

That's the trouble. We just don't know.

Nor do we know precisely how much of a threat to us humans it might be. There have been fatalities, certainly: once infected, the mortality rate is as high as 50 percent – much worse than Covid, with which the virus is said to share a number of characteristics. Unlike Covid, however, H5N1 has not been shown to move between humans.[16] Direct contact with a diseased bird (or, it is starting to appear, infected animals, usually predators or scavengers of diseased birds) is required. Considering how widespread and virulent the disease has become, human deaths are few. For our own sake as well as the terns', we can only hope the colony is spared. If it isn't, the RSPB protocols are quite terrifying: full hazmat gear, no ringing to be undertaken, corpses to be disposed of with great care and above all safely. Government action is aimed very much at domestic poultry (keep indoors and cull if it becomes necessary) but its advice on what to do about wild birds suspected of having the disease is understandably less strategic and less helpful: if a set number of particular victims is found in one area, then they should be reported and might then be tested by Defra; dead wild birds must not be touched and if found on public land need removing by the council, on private land by landowners or their contractors, and so on.

At the moment, given that we have seen no clear evidence of the virus we will proceed as normal, taking due precautions and keeping a sharp eye out. What else can we do?

The last day of June coincided with the incubation period of our first tern eggs reaching 18 days. They will hatch any day now, so we begin our full 24/7 wardening by starting night-shifts. The bad news is that Mick, our most

---

16 Some of our most familiar diseases – mumps, chicken pox, rubella – jumped from animals in the distant past, and the possibility remains ever present. Some epidemiologists were expecting bird 'flu, not Covid, to be the next.

experienced night-watcher, has been forced to withdraw from the team for personal reasons. He has effectively mentored younger volunteers and wardens throughout the life of the project. We have our formal guidance protocols for night-watchers, of course, but here's some of his practical advice to a night-shift virgin:

Top tips for nights.

Don't get cold, don't get wet. Take a sleeping bag or blanket if you can: it gets pretty nippy down there.

Most predators hunt, as a general rule, a couple of hours from dusk and a couple of hours before dawn. Ringed Plovers and Little Terns will let you know if something is about. Ignore the Oycs unless it's a persistent alarm (they're a bunch of noisy bastards, and can be at it all night). You will soon get to know whose shiny eyes are whose. Fox is always on the move. If you get a torch on it, keep on it – it doesn't like it. And a loud shout ('Gerrout of it yer ******* thing!', I find effective) along with the bright light usually does the trick. Same with owls, though the chances of getting on to one are slim.

...I've never seen anyone else down [at the Ponds] after dark all the years I've done it, but if you do feel uncomfortable, pack it in.

Shutting your eyes for an hour or two is not a sackable offence (though nobody has been caught doing it to test that theory).

Hope this is a help. Have a good night and don't dwell on the possibility of flesh-eating zombies or vampires turning up....

He's a loss. The arrival of Ben, our third residential volunteer, offers some cover, but he is here only for the month and it seems inevitable that the pressure on James, Rob and the other long-term volunteers will increase. At least those picking up the night-shift slack will have the benefit of exciting new (to us) technology: this year we begin trialling the use of a thermal-imaging telescope, kindly supplied by Natural England. In return we will provide them with data to contribute to their own investigation into nocturnal predator incursions and the different alarm behaviours prompted by them. Night-time wardening normally relies heavily, if not exclusively, on hearing; however quickly we then respond to the birds' alarm calls (by shining a torch into the colony), we're starting at a

The thermal imaging telescope gives us an advantage over night-time predators. Here a Fox is spotted at the north end of the site...

...while other nocturnal visitors become visible too.

disadvantage – we're already a little behind the game. The thermal 'scope allows more proactive vigilance. We get a better sense of what's happening generally, can to some degree see trouble coming and can get a fix on any intruder more quickly. The results are immediately a revelation: Foxes are present pretty much every night; up to three Badgers have been seen, along with a Weasel – or Stoat – that would almost certainly otherwise have gone undetected. Roe Deer are also quite regular and it is clear that the terns are no fans of Hares in the colony (two more leverets have just appeared), alarming vociferously.

Thankfully, it is only the Hares that we are seeing actually inside the colony: the fence, recently extended down to the receding edge of the Ponds, is doing its job, with little more than the occasional hiccup. And we need it to: we have our first Little Tern chick (LT1's), hatched on the 5th. In less than a week there are at least 25, and five new nests have also been found, bringing the total to 35. How many are active is not entirely obvious: James and Rob reckon 29, though the mobility of the chicks puts accurate counting at a premium (heat haze doesn't help either, and it's been very warm recently, despite a sometimes chill wind). For the first two or three days the newly hatched chicks remain in the nest, digging their feet into the substrate (if you weigh only seven grammes it's as

A Little Tern chick doing its best to be unobtrusive. We check very carefully on our walk-throughs, but inevitably miss such well-camouflaged birds.

well to hang on), brooded by their parents, who will hover angrily and noisily above – well, anyone or anything that strays too near, however blameless they might be. After that, chicks leave the nest, scatter and rely on their excellent camouflage to keep safe – "as hidden as a thought unborn", as Clare has it.[17] They might seek a little cover by hunkering down in a small hollow; our wardens have reported seeing adults make scrapes for just this reason. They might also seek shelter from the weather by crouching in the lee of plants, jetsam or even human rubbish; or in one of our purpose-built little shelters – we've continued to provide them since our first experiments nearly two decades ago. Once the chicks are away from the nest, their parents only brood in bad weather, and then only until the chicks should be able to look after themselves. They do, however, keep careful watch, leading chicks to nearby cover (in our case, the vegetated dunes) and hovering over threats on the ground. The whole colony will go up in response to an aerial predator. The chicks themselves, already showing through their down the pins of what will become their primary flight feathers, will scatter at improbable speeds and with remarkable agility; they might end up anywhere, needing to be relocated by their parents. The older they get, the further they scatter.

We local volunteers have been mostly filling the four-hour evening shifts. I'm glad. The colony from the late afternoon, through twilight and into the night has a particularly powerful ambience, and an ever-more intense sense of place as the slowly sinking sun's light continuously reshapes it. When the huge skies are clear, the sunsets are spectacular, almost, it seems, as a matter of course. When the weather is more mixed – you can see it coming from miles away – the effects can be even more dramatic. I have seen high-flying terns showing as no more than specks of light against lowering, purple-grey clouds; underlit by the setting sun, they look like diamond dust. This evening all is clear and still. It's busy enough, with birds coming in off the sea, lots of waders fiddling around, and a big mixed flock of Black-headed and Little Gulls (a favourite of mine) hawking for flies over Long Bank Marsh to the west. Sandwich Tern numbers are growing, too, but I can see nothing in distress, and certainly nothing dead or dying.

The terns themselves are busy bringing fish in for the youngsters: sand eels, mainly, but also small flatfish, clupeids and crustaceans. The Ponds sometimes

---

17 Clare's actually talking about the Nightingale's nest (in his poem of that name), but as a description of natural crypsis generally I can think of none better.

Fishing the Ponds: a Stickleback is plucked from the water.

come into play, as does the Humber, but mostly when marine conditions are too rough to fish;[18] generally most of the fishing goes on out to sea. And often not that far out: when the tide turns and prey begin to emerge, the terns are waiting for them. It's quite terrifying to watch the birds plunge-dive from ten metres into what can only be a few centimetres of water. The Rev C. A. Johns (1811–1874), yet another prolific Victorian parson-naturalist, describes the Little Tern's hunting technique in his *British Birds in their Haunts* of 1862:

> it is seen flying slowly along, some fifteen or twenty feet above the surface of a shallow tidal pool, or pond, in a salt marsh. Suddenly it arrests its onward progress, soars like a Kestrel for a second or two, with its beak pointed downwards. It has descried a shrimp, or small fish, and this is its way of taking aim. Employing the mechanism with which its Creator has provided it, it throws out of gear its apparatus of feathers and air-tubes, and falls like a plummet into the water, with a splash that sends circle after circle to the shore; and, in an instant, having captured and swallowed its petty booty, returns to its aerial watch-post.

---

18 Even so, sometimes there is no sanctuary: in 1997, 30 chicks were lost to starvation when storms not only made the sea unfishable but also chilled the birds with heavy rain and buried their nests under gale-blown sand.

Johns's writing, and his understanding, are very much of their period, and as a great populariser of natural history he was not always regarded very seriously, but here he seems to me to capture the moment of the action – "it throws out of gear its apparatus" – at least as well as many more modern texts do.

The terns' success depends on the health of the sea and the condition of prey populations, and at what point in their own breeding cycles those prey species are. Quite how the parents find their young in the colony is not fully understood. They recognise each other's calls, the young probably learning what its parents sound like while still in the egg, but even so it's not a straight-forward business: any seabird colony is a cacophonous place. There's some

Fishing the sea: a helpfully calm day.

evidence that adults call as they fly around over a crowded beach deliberately to see which chick consistently follows them on the ground. But this is not a perfect system; not only must a hungry chick avoid careering straight into the bill of an opportunistic Oystercatcher, it must also be careful not to approach the wrong parent. If it does so it is liable to meet with aggression. These little dramas are often over quickly and might even go largely unnoticed, but on occasion they can develop into much longer and relentlessly unpleasant confrontations. On the other hand, I've just seen five small chicks gather begging under a hovering fish-bearing adult like party children demanding sweets from a harassed grown-up, and on this occasion the adult simply gave up, flew off and ate the fish itself. It's a complicated process, feeding. Nonetheless, when conditions are good – when the weather is not too hot, too cold or too wet, the wind not too strong, predation light, the tides low and human disturbance controlled – a chick after ten or twelve days will have lost its egg tooth, quadrupled its body weight, replaced much of its down with feathers and doubled the size of its bill. It will start to look something like a Little Tern.

There's a sudden kerfuffle over the dunes just to the north: several adults are repeatedly mobbing… what? Try as I might I can see no threat. This is not altogether uncommon: the same thing happened to James in another area (over LT1's empty nest) only a couple of days ago. Perhaps a leveret, a Weasel, a Grass Snake, or even a small rodent, moving through hollows and behind bumps or vegetation? Or something that is simply new, or out-of-place, just not right? Whatever it was, or wasn't, the adults eventually calm down and drift away.

As the light slowly drains, the rhythm of the colony slows and the birds quieten, settling into their evening routines of bathing, preening or just loafing around. Some are dozing, though heads come up quickly enough when the first of four Herons drops on to the south shore. These are this year's young and we always watch them carefully; Herons are included on our list of potential predators, though we've never witnessed a kill. The terns and plovers treat them with careful circumspection but seldom go into full alarm mode. The same is usually true of all, bar one or two of the larger gulls – the Black-backs and Herrings – which fly over regularly: most of the time they are simply to be kept an eye on. Their potential prey seems able to read intent quickly and accurately, to know when a threat is likely to materialise. For now, things are quiet, idyllic, even.

Of course, shifts are by no means always so enjoyable or pleasant. Sometimes they are just plain boring: Bill Oddie's dictum that "there is always something" to look at may well be true – a closer examination of a bird you've taken for granted, a bit of unusual behaviour – but it's not always easy to hold on to, especially when you are stuck in a hut in poor weather with reduced visibility. Nobody else is daft enough to be out, so there are no walkers on the beach to talk to or argue with and, like you, the birds are hunkered down. It's too rough for them to fish, and there are eggs or young to protect. I'm too tall to get really comfortable in our huts, so those particular pleasures of cosiness in the face of bad weather don't come into it. The huts are draughty at the best of times, but in high winds they can be unsettlingly rickety, and the violence of storm-driven surf can sound horribly close. Boredom is your enemy, and you might wish for some excitement – though it is probably not going to be of a welcome kind. So you have to try to say alert. Predator attack may be unlikely in the conditions, but if it happens, it happens with alarming speed. The greater problem might be a high tide, driven higher by the wind, threatening nests. Losing concentration can be disastrous (it has happened). I try to keep focused by writing (this paragraph was drafted as the rain bashed down) or listening to podcasts. I can't read: I find I get too absorbed.

But for now those bad weather shifts seem a world away. I allow myself to relax a little and take in what else is happening. The Shelduck family, preening on the far shore, seems to be down to seven chicks now. Still not a bad performance, if they can hang on to them. The likely villain is Fox, though we have only the very

Rain drives in from the north – and the photographer is
about as far away from shelter as he can get…

slender evidence of seeing one pounce on *something* small some distance away from the family. A flotilla of half-a-dozen non-breeding male Mallards cruises the open water, full of fairground machismo despite their failure to pair up. Little Grebes pop up and down all over the place: it's almost impossible to get an accurate count. More and more Little Egrets come in to line the far shore. Around them, another colony is fast developing: Hart's-tongue Fern seems to love growing on the stony ground and proliferates there, inhospitable as it seems: it must be getting its roots down into the nutritious clay beneath. On the breeding shore – the side where I'm presently slumped semi-comfortably in my chair – there is ever more vegetation getting hold. I'm a poor botanist, but even through my binoculars I can identify a number of the colonisers in this habitat. I follow the advice of a friend who *is* a good botanist: decide what other, familiar plant your plant resembles, then stick 'sea' in front of it and you'll have a good chance of being right. So: Sea Bindweed, Sea Rocket, Sea Holly, Sea Mayweed… Soon the Sea Asters will start to flower; they do well in the area, and drifts of these yolk-centred, pale lilac flowers are a real feature of the late summer and early autumn. Otherwise, Common Restharrow similarly isn't quite there yet but will put on its own pink show soon. In the dunes, there is Wild Carrot, Ragwort, Bull (Spear) Thistle and Perennial Sow-thistle and, in the wetter areas, stands of Common Reed. But above all there are grasses: Marram, Sea Couch (and maybe Sand Couch – here my ID skills reach their limit), Lyme-grass and Dune Fescue, and it is these which trouble us most, even more than the Sea-buckthorn. Perhaps the grazing Hares keep them back a little (although the peppery Sea Rocket might be more attractive), but if they do, they're not doing a great job. These are tough, tough plants. They need to be, as theirs is an astonishingly harsh environment for a plant to survive in: a motile, low-fertility growing medium, lashing, dehydrating winds, the threat of inundation by salt water. Their many adaptations include matted and rhizomatous root systems to hold on to the shifting sands and spread as far and as quickly as possible. Marram-grass in particular is quick to colonise and gets a grip even on the top of more newly formed 'white' sand dunes. It is this which makes it so important to our natural coastal defences: its resilience and vigour as a binder of dunes. It's also what makes it a particular problem for us. We need selectively to clear areas in order to keep the tern habitat going, but we'll have to make a strong case to the authorities for them to allow us to do so. We return to the subject frequently in our monthly Management Committee meetings (where both the Environment Agency and Natural England are represented), and the issue is becoming ever more pressing. We need to act soon.

Our July meeting gets a 'State of the Union' overview of the colony's progress from James and Rob. The weekly report it's based on (even with my editing) gives a good idea of how thorough and detailed the monitoring is:

JW, 13/7/22, 1800–2200

Weather: SW-s f2, CC 4/8-3/8, 17–12C, sunny, calm.

F: 7.9

T: 11.9

N(et): 7.2

Little Terns: Max 70 @ 1900…very mobile.

LT 24, 25, 27, 28, 30, 31, 32, 33, 34, 35, 36, 37 all sitting on nests.

Birds not on nests as yesterday: LT1, 3, 5, 6, 7, 8, 9, 11, 15, 16, 19, 20, 21, 23, 26.

New birds not on today: LT12, 22, 29.

Newly hatched: none.

Chicks: 3 seen from VP, 2 near LT37, 1 by LT23. 2 seen from New Hut, 2 near tall post by LT15. One left of RP7 not seen but birds dropping in and out of sight with food.

New nest LT38 to left of LT21… seen sitting since 11/7 but probably been sitting longer than that… Can only see from outside and left of Hut.

Ringed Plovers: RP14, 15, 17, 18, 19, 21, 22 sitting. RP20 reportedly hatched, didn't see. Amazingly no chicks seen but couldn't look for long.

Oycs: OC4 1 chick by raft, OC9 sitting, OC8, re-lay.

BhG: BH1 fledgie on raft again.

Predators

No Kestrels again… .

And this is just one, four-hour shift. To get a full, week-by-week account of the state of the colony requires all shift reports to be put together. It's a big job.

James notes the absence of Kestrels during his shift precisely because they have become an increasing problem that we have had to address over the last couple of weeks. Although in general terms they are recognised, along with Fox, as a top predator of Little Tern chicks, they have not been a significant source of concern for us in recent years. Even in the more distant past the problem they have posed has usually been intermittent rather than regular; in the mid-1990s, it was Merlins that were the greater threat, with three or four birds regularly visiting the colony over a period of four years. Still, when we have had a Kestrel issue, it has been a major one: in the 2001 season, a female visited as many as 20 times a day, taking most of the 70 chicks present. The impact can be colony-threatening: in 2001 (again), they took over 500 chicks from the Great Yarmouth colony.[19] More recently, they wiped out the colony at Chesil Beach in Dorset, taking at least 100 chicks; it has taken until this season for that colony to regain lost ground.

This is not so much predation, nor even scavenging, as harvesting.

In our case, the Kestrels in question come from a nest at a cottage to the south – one owned, ironically enough, by one of our volunteers, Marcus. The breeding pair has visited the colony throughout the season without showing real intent, but from the first week of July both the number of visits and sharpness of purpose have increased dramatically: 13 visits on one day, five on the next, five again on the day after that. While most sorties come to nothing, chicks are being taken, and as the Kestrels themselves have hungry mouths to feed with three of their own chicks (one of which is close to fledging), it is unlikely the attacks will stop of their own accord. If Chesil's experience is anything to go by, the impact could end up being serious indeed. Preventative measures so far consist of tying rags to long bamboo canes and waving them furiously at the approaching birds, having studied their regular flight paths into the colony. It is effective, to a point, especially if accompanied by shouting. Another option to try is air-horns. James used these to good effect in Canada when he was

---

19 Kestrel predation at the Great Yarmouth colony was monitored over a period of 17 years. In that time at least 3,436 Little Tern chicks were taken – compared to 2,536 chicks fledged.

Our troublesome male Kestrel hovering over the colony.

working with… bears. Surely if you can frighten off a bear, a Kestrel or two shouldn't be a problem?

We'll see. We agree to order a couple of horns, express delivery.

Of course, the difficulty is that in trying to repel the predators we are also disturbing their prey more, but there is nothing else for it. At least the terns and the other shore-nesters are used to our presence around the colony. But a happier solution presents itself: diversionary feeding. We have tried this before with Foxes, picking up road-kill and moving it near – but not too near – to the colony. (One heroic warden once dragged a rotting Seal carcass up the beach for the same purpose.) The aim is to provide an easier meal, and it does work. The dangers are that you attract more scavengers to the area, increasing the exposure of the colony, and also further test the patience of human beach users, so where exactly you put the diversionary food source is a matter of fine judgement. With Kestrels, the strategy is easier to deploy: studies have shown that frozen surplus hen chicks, left within easy reach, are an irresistibly convenient food supply for parent birds trying to feed their young. Tern and other shore-nester losses have been dramatically reduced by employing this method. Even so, trying the scheme out can prove problematic. One colony had its Kestrel-feeder station vandalised, because, it seems, members of the public thought someone was trying to poison the birds.

So it's fortunate for us that the Kestrel nest is in our friend's garden: Marcus is happy to allow us access, and even to use his freezer for the frozen chicks. It's important to stop providing food as soon as the young Kestrels have fledged, lest they become over-dependent, but with a little luck, local farmers will soon start harvesting their fields, and the resulting displacement of tasty small rodents has in the past taken the pressure off the colony.

We agree to try.

There's a wider irony in all this. As anyone who feeds their garden birds knows, the Robins, tits and finches you attract themselves attract their own predators: notably, Sparrowhawk. It's a feeding-station for them too, and not everyone is pleased to see them. They may be more pleased to witness acrobatic Grey Squirrels taking advantage of their peanuts, but those same squirrels, possibly supplied by many feeders in a small area, might well as a result maintain a higher than normal population. And Grey Squirrels also feed on songbird eggs and nestlings – as the birds in my own garden know to their cost.

Any nature reserve or bird colony is in effect a huge bird-table when it is located in an area substantially stripped of natural food resources by human development of one kind or another. My garden, in a small village, is a nature reserve within a larger nature reserve of woodland and more gardens (often with bird feeders), surrounded by many square kilometres of agricultural prairie. So my approach to conserving birdlife contains a paradox: at home, I want to deter Grey Squirrels, but welcome and support Fox, Badger and Hedgehog – the very creatures we expend time, energy and money in keeping out of the Little Tern colony.[20] (Thankfully Grey Squirrels are not a problem there.[21]) In its own small way my situation reflects a wider truth about conservation: any intervention we make – the priorities we choose – on behalf of at-risk species can have consequences,

---

20 The predation problems at the Chesil colony arise from Kestrel (mainly) and Hedgehog. Both species are in decline, the latter quite calamitously so.

21 Not *quite* true: there is one historical record, from the 1990s, of a Grey Squirrel at the colony. There is no indication it did any damage; but then, a less squirrelly place is hard to imagine.

sometimes negative, for others. That we have to make the kinds of choice we do is the essence of the management role – the stewardship – we have had to take on, the balance we have to try to keep.

Our diversionary feeding of the Kestrels works immediately. For several days their visits to the colony stop altogether.

And then their chicks fledge, we stop the feeding… and they reappear.

Usually it's both parents and sometimes one of the chicks. For the remainder of the month they make several visits a day, occasionally up to a dozen. They aren't doing a lot of damage, but our flag-waving and air-horn blowing (though opinion on the efficacy of the latter is divided) seems at least to be knocking them off their stride. We have the advantage over larger colonies such as Chesil because we can police more intensively, but there's still a price to pay: we have to concentrate our wardening at the southern end of the colony because that is where we can spot and intercept the Kestrels as they come north from the nest-site, but that means less time monitoring and recording from the Viewpoint: we simply can't get around to the opposite shore fast enough when danger looms. The increasing numbers and mobility of both tern and plover chicks, ranging from the newly hatched to the fully fledged, makes consistently accurate counting very difficult even from the right side of the Ponds: from across the water it's almost impossible.

Counting is also complicated by heat haze: it's been very hot across the country for the last few days, with temperatures touching the high 30s and even higher in one or two places. Drought threatens. The papers carry stories of dehydrated Hedgehogs, Fox cubs and nestlings and there are reports of Swifts falling out of the sky in London. An RSPB helpline is struggling to deal with the surge in calls. Wildfires have broken out in a number of places, wreaking havoc throughout food chains. Conditions are less severe at the Ponds thanks to the cooling sea breezes, but it's still hot, and that threatens young birds. Nor is the sea itself immune: the heat also impacts on the marine environment, especially on static species such as barnacles, mussels and sponges. Sea temperatures are already rising, of course, pushing colder-water species further north. Again, the effects are felt from bottom to top of the food chain.

Despite that, there is good news: we have our first fliers! Wardens spotted three take to the air. Young birds have been seen exercising their wings for a little while now: we record them, inevitably, as 'flappers' (a term also used by shooters for young duck and partridge). These first flights are short and not entirely gainly, but it's big landmark. We are also confident we have 35 nesting pairs of terns, we think 31 chicks altogether (though our maximum count couldn't find more than 18 in one sitting), but we know we have lost one of the three remaining nests on the beach. The nest we thought we had moved beyond the reach of the sea has failed, probably because of a tidal surge. Sadly, the mother (it's not clear to what extent if any the male shares incubation) has continued sitting on a lifeless egg for 39 days: it should have hatched nearly three weeks ago. We know we've had 25 Ringed Plover chicks hatch, but can see only 15, five of which should, by date of hatching, have fledged; two more are only now on the point of hatching. They are doing better than appeared likely earlier in the season.

As July wears on the colony becomes busier than ever, with the numbers of visiting birds, both adults and this year's young, building up all the time. Passage waders include Whimbrel and Knot, with the latter including birds ringed in Norway and the Netherlands. But Black-headed Gulls dominate, and the numbers of their young mean we have lost track, temporarily we hope, of our own fledgling. Common Terns and Arctic Terns are increasingly present too: one party of seven of the latter included two juveniles. These are birds that almost inevitably will have come from or through infected colonies elsewhere. But although Rob has taken the precaution of removing one dead Sandwich Tern from the colony, there is still no clear evidence of bird 'flu being present. (Our eldest Oystercatcher chick succumbed, possibly to the heatwave, but that is the only other fatality recorded.) And there is, as it were, plenty of opportunity for the virus to get hold. Quite apart from the gulls and other terns, there was a maximum count of 254 Sandwich Terns (including 95 juveniles) on the Ponds' edge one evening. Eight were sporting colour rings on their legs, giving us valuable insight into how these birds move: four had been ringed in Norfolk, two in Scotland, one in the Netherlands and one in Spain. We'll have to chase down further details of the ringing sites. There have also been two fledged Little Terns we believe have come from elsewhere, and a further two birds were seen flying south. In fact, it's clear that Little Terns – adults, young and non-breeders – are starting to move along with their congeners, and again there's a slightly unusual number of inland records: 13 in Cambridgeshire, nine at Fairburn Ings on the North/West Yorkshire border, six from South Yorkshire, five in Lincolnshire and singles elsewhere.

With all this activity, keeping the Kestrels at bay as best we can – at least until harvesting kicks in – is likely to become the key focus of the rest of the season. Not that they are the only predators around: Weasel and Fox remain a threat (the Shelduck family is down to five, though neither suspect is bang-to-rights), but aerial predation is the greater issue, and not just the Kestrels: Barn Owl and Marsh Harrier are more frequent but have so far not been seen to make any attempt on the colony; Sparrowhawk, Peregrine and Hobby visit mainly at high tide, making the roosting wader flocks the likely target. An Arctic Skua hung around briefly, but its interest lay in stealing fish from the adult terns rather than taking their chicks. Crows lurk but seem largely uninterested. Aside from the Kestrels, whose visits are not diminishing in number though they seem to be enjoying little success, it is the Sparrowhawk which is most dangerous: he (it is a he) adapts his hedge-hunting tactics to the beach environment, flying in low from the south along the seaward side of the dunes then flipping through or over them into the colony in the hope of taking an unprepared bird. He's had some success, mainly (inevitably…?) with the plovers.

Overall, however, predation levels are low. This is certainly down in part to our wardening, but it's also the case that the very busyness of the colony can make life more difficult for predators, offering a confusing plethora of targets many

A Ringed Plover chick legs it.

of which – the gulls and the larger terns especially – are aggressive defenders of their air-space and indirectly of the smaller terns, plovers and other waders.

So the month ends strongly. A walk-through to ring chicks helped enable us to get a more comprehensive view of the state-of-play: we have 31 tern chicks, 24 of which are believed to have taken their first flights. Ringed Plovers are hanging on with 14 chicks, with more nests still to hatch, and three Oystercatcher chicks (from two pairs) are thriving. Up to 5,000 Dunlin and 2,000 Knot are now coming in to roost, and human disturbance, mainly in the shape of dog-walkers and photographers, is low, though a low-flying aircraft caused a minor panic one day.[22]

Most spectacular, however, was the gathering witnessed by our local volunteer Georgia of no fewer than 205 Little Terns on the Ponds' edge in the evening of the 29th. This was clearly indicative of a large movement: the highest count for the week of our own birds was 116. Typically, the visitors had all departed by morning. Two-hundred and five is our highest confirmed count of Little Terns at the site in 80 years.

It's all go.

---

22 Perhaps the least expected and most disturbing event, however, was witnessed by our volunteer Ben, who stumbled across a couple very much *in flagrante* in the dunes. Ben, as planned, left us at the end of the month, taking our thanks and some mental scarring with him.

AUGUST

Birds don't really do summer, Little Terns least of all.

By the end of July and the first week of August it is clear that early-breeding adults and young are already starting to leave the colony: there's a big exodus on the 5th, and 90 birds gather briefly at the Breach on the evening of the 6th. These are not all ours, I'm sure, though only 20 birds were counted still in the colony at the same time. The day before, Gronant tweeted that it had seen its second-best ever fledging numbers (209), and that the site was now officially 'down'. At the same time, Winterton reported that its last birds had left, with them having seen around 300 breeding pairs produce at least 585 fledglings, and possibly significantly more. We still have birds, but are seeing our numbers fluctuate considerably, as fledglings and adults from other colonies drop in: one day saw a maximum count of 132. Based on careful analysis of plumage states against known laying and hatching dates (this is where the value of numbering nests really makes itself felt), James and Rob reckon that 29 of 38 fledgling-sized birds might have been ours – and they all leave the colony a couple of days later.

Fledglings among the adult birds, now outside our protective fence.

But unlike Gronant and the other colonies, we still have chicks, and even eggs. We count 18 chicks, some little more than a week old, and two nests still to hatch. Most birds have now moved down almost to the southern tip of the Ponds, outside the fenced areas, and are favouring the low islands and spits that have emerged as the water level has fallen. Ringed Plover chicks (14, and four fledglings) are still spread around the colony, and at least one bird is still sitting. We're down to just the one Oystercatcher fledgling.

Although visiting birds continue irregularly and briefly to add to our home population the overall trend is clearly downwards, and that brings an un-looked-for complication. We had hoped and expected that harvest-time would see our two main predators – Kestrel and Fox – switch their attention away from the colony and towards the displaced rodent populations in nearby fields, as they have in the past. And indeed the harvested fields are now attracting the attention of many Kestrels: we can see any number of them hovering away in the distance.

But the adult male from our local nest has, it seems, developed a taste for Little Terns and their chicks, enough to make him forgo the pleasures of juicier voles. Although he is not alone in visiting the colony, he is by far the most regular and effective hunter. Kestrels have made over 40 visits across the first week of the month alone, and while we can only be reasonably sure of two kills (one large tern chick and one plover chick) we strongly suspect several more. The second week of the month paradoxically sees the number of visits drop sharply to seven but the confirmed success rate rise, with at least four chicks being taken. Another two are lost the following week. What seems to have happened is that the male has adapted his tactics, prompted perhaps by the move of his prey to the more open area outside the fence to the south, and perhaps by our own doubtless irritating interventions; whatever the reason, he is now coming in fast and low, like a Sparrowhawk (also visiting regularly), so that we see him coming too late. We simply can't move fast enough to block or distract him. And, of course, the diminishing population of birds in and around the colony means fewer eyes to spot danger coming and fewer numbers to mob intruders: the remaining birds see him too late, as well. By mastering the technique he has effectively created a personal feeding niche without significant competition from other Kestrels. These tactics have been observed at other colonies, but the last time they were recorded at the Ponds was over 20 years ago.

Fox visits increase, too, as August wears on; one finds its way inside the fence for the first time, and one was seen to take a tern chick from the edge of the water mid-month. Crows have taken a couple of hatchlings. But the real problem has been the Kestrel.

MH, 09/08/22, 0600–1400

NE f1, cc 2/8

F: 7.4

T: 9.2

N: 5.5

LTs: max of 59 @0611, with numbers dropping off to c 10–20, then fluctuating between 20 & 30.

LT39 sitting.

Fledglings & Chicks: 1 well-developed fledgie by Ponds' edge with 2 fledgies, 2 large chicks (2 med/large chicks. 1 med/large chick behind tern table inside fence, 1 large chick behind water raft just within fence, 1 small chick just within fence to left of beach raft, 2 large chicks in LT22 area (1 was U32), 2 small chicks near south end of net fence on beach…

RPs: RP23 sitting until 13.00 when it disappeared – still 2 eggs visible through scope. Fledgies: 1. Chicks: 6 […]

Oycs: 1 Fledgie.

Shelducks: 5 chicks

Predators:

Crow: landed inside fence near S Hide being mobbed by terns. Unresponsive to shouting and flag-waving so I ran up, jumped the fence and scared it off west. So far as I could see it hadn't killed any chicks.

Kestrel came from S @ 09.46. Scared it off with horn and flags. NOTHING TAKEN.

Hobby fairly high over S Hut and W @ 12.09. NOTHING TAKEN.

Human disturbance:

People 6 N, 10 S.

Dogs: 2 N, 1 S.

Battery change by YWT – 3 old for 2 new.

Other:

8 Little Grebe, 1 Black-tailed Godwit on S end of Ponds, 1 Greenshank on N end, 8 Common Terns over W, 6 Cormorants (1 on Ponds 5 N), 7 Whimbrel SE, 1 juv Ruff

A typically thorough 'state-of-play' report from Murphy, who is soon to leave us to complete his BSc in Biological Sciences with Ecology at Edinburgh University. Two more residential volunteers, Beth and Zach, will provide cover to see us through the last couple of weeks.

Other predators seen in the area include Hobby, Marsh Harrier, two Peregrines, Buzzard, Red Kite and a Stoat, but we don't know if they've met with any success. A juvenile Harrier stooped on the Shelduck chicks, but the parents saw it off successfully. But end-of-season mortality isn't just down to predators. Some birds have simply started too late to give their chicks a decent shot, and they die in the nest or soon after leaving it. Other parents might abandon their efforts at the egg stage. Even older, well-developed chicks are not safe; we have found a dead Oystercatcher fledgling, the one that we had ringed on our first walk-though. There is no obvious cause of death, but it's the second such bird we have lost this summer, and it follows a similar incident last season. We can only assume dehydration at the end of a hot summer, though why this should strike just as the birds are achieving maturity is a mystery. Again, there is no evidence that bird 'flu is responsible.

And then there are the tragi-comic breeding attempts that never got off the ground; the Little Tern found to have been sitting on a snail shell... the Oystercatcher nest where the adult had been brooding a tennis ball. We know this because we had our last walk-through of the season on the 8th of the

A tern chick, only a few days old, getting its metal
ring during our late July walk-through.

month, when we ringed five more Little Tern chicks and one last Ringed Plover chick to add to the ten ringed at the end of July.

Ringing breeding birds is always a finely judged business, and its organisation and implementation is in the experienced hands of Obs Head Warden Paul Collins. The window of opportunity is small: the chicks need to be ringed in the first few days of their lives, when they are still tied to their nests.

Once they are fully mobile, their parents' alarms as we approach send most of them scurrying at speed into the cover of the dune vegetation, where it is simply too risky to follow them. And our scheduled walk-throughs will unquestioningly be postponed if Paul judges that the colony is already under too much stress – from disturbance, predators or bad weather – or generally in a weakened state from low productivity, malnutrition, dehydration or other factors. So, for instance, no chicks at all were ringed in 2016 as their numbers were so low. In good years, between 30 and 50 might be ringed. This year the possibility of bird 'flu has been an added complication. Some reserves have decided not to ring at all but, given the lack of concrete evidence at the Ponds, we have decided after much debate to continue – with extra precautions: fewer and shorter visits, disinfected hands and footwear and sanitised clothing.

There is one exception to our policy: a couple of years ago we started exploring the possibility of trapping and ringing adult Little Terns. The technique, using a remotely controlled spring-trap, has been used with great success at the Gronant colony – where 150 birds had been trapped in just three years. (The trap is essentially a large, square, flat frame with a rolled net spring-loaded along one side. It is placed on the ground over a nest and when the adult returns a hidden watcher springs the trap remotely, throwing the net harmlessly over the bird.) After extensive trials in the Obs garden, our trap was finally tried on a wild bird in the colony: the ever-put-upon Ringed Plover. It worked, simply and straightforwardly enough to encourage us to attempt it on a tern. Kieran, our project leader, picked a nest-site on the edge of the colony – to minimise disturbance – and deployed the trap. Again, it worked without complication. Not only was this the first adult Little Tern to be trapped at the Ponds, it was also the first 'control': that is to say, remarkably, the bird was already bearing a ring. This was enough to tell us, after a little research, that the bird had been ringed as a chick at the Winterton colony in Norfolk in 2013.

Unfortunately, bad weather meant we decided not to continue with the project last year. And our wish to minimise disturbance even more than usual because of bird 'flu means we are postponing again this season. It's frustrating, but undeniably for the best.

Why expend so much effort and energy on ringing at all? It does have its opponents, who see it as an unwarranted intrusion into wild lives. And it is certainly the case that the return on that effort and energy is very low: across *all* ringing schemes the average recovery rate is 1 percent. But despite technological advances with tagging and satellite tracking, it remains for the time being at least the only real way we have of getting a clearer understanding of the movements of wild birds and as such is invaluable for their conservation.

Migrating birds – tens of thousands of them – have been ringed along the length of the peninsula for over 100 years: Spurn is one of the key sites (many would say *the* key site) for bird migration in the British Isles. So there is no shortage of experience or expertise. Andy Roadhouse notes in his monumental *Birds of Spurn* that a Little Tern ringed as a chick in July 1914, 30 years before the founding of the Obs, was recovered in Portugal the following September.

Between 1945 and 1985, 77 birds were ringed by the Obs, though there was only one recovery, and that was within the British Isles. Ringing has taken place at the Ponds colony since the 1990s, but data collection was standardised only from 2014. Between then and now some 140 chicks or pulli (birds still in down and unable to fly) have been ringed. Rings are of three types. A numbered metal ring, the most common, and administered by the BTO, is too small to be read in the field unless under very exceptional circumstances, so 'control' is only possible if a dead, dying or trapped bird is recovered. These are the only rings that can be fitted to very small chicks. Coloured rings, easily visible in the field but giving only the site of ringing (though a BTO ring might also be present) were introduced some years ago: these could be fitted to larger chicks, but as they bore no additional information – such as the year – proved a limited success. The bird still had to be trapped, or found dead, in order to obtain that data from the BTO ring. Darvic rings (the name derives from the PVC sheeting employed to make them), both coloured and numbered, are more effective: provided they can be read in the field, which is often still difficult (Little Terns

Though clearly older than the chick pictured on page 96, this bird was in fact ringed (with a darvic) two weeks earlier. (Because Little Tern eggs are hatched throughout the season, near-adult birds will often be present alongside newly hatched individuals.)

have irritatingly short legs, for one thing), both location of ringing and details of the individual bird are accessible. Which leg the rings are put on gives further information: for chicks, a darvic goes on the left, a metal ring on the right; for adults, this is reversed. (Beacon Ponds and Spurn darvics are yellow with black letters and numbers, U01–99.) Where sufficient detail can be read, a report is sent to a central database, and so patterns of movement and other behaviour can be built up.

It must be stressed that the presence of rings causes birds no inconvenience at all – other than that of the original fitting, of course, and in experienced hands that is minimal.

To hold a wild bird in the hand is a special experience. For most of us the opportunity comes only as the result of life-threatening misadventure: a garden bird rescued from a cat or from entanglement in pea netting, a Heron that has flown into a powerline, a Pheasant hit by a car. I recently picked up a Barn Owl, a traffic victim, from a gutter near home, hopelessly broken. I thought it was dead, but it turned its head enough to meet my eyes and feebly tried to rake my wrists with razor talons. Normally it would open my skin easily, but here it was so weak it couldn't even mark me. On the very point of death, hopelessly, it was still trying to preserve its life. To look into human eyes at a similar moment would have been unbearable. The owl's eyes remained as blank, as unreadable, as a bird's eyes always seem. Yet I was still moved to tears, almost wishing it *had* drawn blood in one last gesture of defiance.

So I'm acutely aware of the privilege of holding and (though it does not know it) helping a healthy wild bird. I am not a trained ringer myself – it's a skilled job with a lengthy, carefully graduated training regime – but I do help out where I can. At the Obs, I have held, ready for ringing by others, thrushes, warblers and flycatchers; at the colony, Oystercatcher, Ringed Plover and Little Tern chicks. It's only when a bird is in the hand that you realise how fragile it is, how small (they are always smaller than you think they are going to be), how light, how… bony: a graceless bundle of feathers, the merest scrap of life. That such a thing will ever fly – and in the case of migrant birds, fly so prodigiously far – seems, for the moment, quite impossible. And then, for adult birds, the ecstatic, transformative moment of release: a moment of natural magic, long recognised and exploited by stage magicians.

The euphoria is certainly mine; but is it the bird's too?

I have on my desk a photograph from last season of a Little Tern pullus, perhaps ten days old, its neck and head protruding from my rather grubby loose fist. I'm about to pass it to Paul for a darvic to be fitted. It is busy violently pecking at my thumb – reasonably enough. My granddaughter has declared it 'cute' and demanded reassurances that it was not being hurt. Of course she's anthropomorphising a wild creature, but how else can we start to try to comprehend other species than by comparing and contrasting our lives to theirs? For myself, at the time I wanted to applaud its courage in resisting me, but is that what it was? Courage? Is that what I saw in the dying Barn Owl? Or blind survival instinct? Where is the line between? Courage, perhaps, though not necessarily in the same way we would understand it. Because another part of the intimacy of the ringing process is facing the inadequacy of our understanding of what it is to be a bird. How do we even know that we are asking the right questions? What questions *aren't* we asking? Do we have the words? Despite the advances in ornithological research – and they are many and extraordinary (see Tim Birkhead's *Bird Sense* and Jennifer Ackerman's *The Genius of Birds*, wonderful books, both) – there is still so much we do not know. As I looked into the young tern's eyes I was reminded of the late Richard Richardson's desire to penetrate beyond the 'inscrutability' of a bird's eye.[23] 'Inscrutability', to which I would add 'implacability'. Certainly nothing human.

And this, it occurs to me, is why conserving other forms of life on our planet is so important: not because they are like us, but because they aren't. Because they encounter and deal with the world in ways we can scarcely imagine, and those ways – sensory, behavioural, experiential – seem to me no less valid than ours. We all live in different worlds, and we all live in the same world.

Lord knows what that young Little Tern 'thought' of me. Very little, I suspect.

So, what has ringing taught us about Little Terns? Specifically, their movements?

Mainly that there is a great deal we still don't know.

The recovery rate for ringed Little Terns is less than 1.5 percent, lowest of the terns generally apart from the Arctic Tern and lower even than the rarest British breeding tern, the Roseate. If you consult the BTO website, you can find a good summary of records, including information about the longest-lived bird

---

23 The revered Norfolk birder and bird artist.

'My' pullus

(27, controlled near Gronant in July 2020 and still going strong at the time of writing) and the greatest distance flown by a bird within the British Isles (400km). At the Ponds, colour rings have confirmed inter-colony movements. Most birds seen over the years (we are talking single-figure sightings each season) have come from the Teesside area (blue), with others from North Lincolnshire (red),[24] Great Yarmouth (yellow), Norfolk (green) and Ireland (different green). Our own colour, mauve, proved unhelpfully prone to bleaching to a useless silvery-white, though we have had some success with metal rings: a chick ringed in 2018 subsequently had a darvic fitted at Foulney in Cumbria and was later seen at Gronant. Another, ringed in 1993, was later found dead at Holme-next-the-Sea in Norfolk in 2016, making it 23 years old (briefly the national longevity record). And another, ringed in June 1999, was later trapped as a breeding adult at Zeebrugge in Belgium in 2004 and then in Heist, also in Belgium, in 2007. One of the first darvic-ringed birds (U04), ringed on 16 July 2014, was seen roosting at Crimdon Dene in Cleveland two years later. And a bird given a darvic as a chick in 2020 has was recovered in the Netherlands on the island of Zuiderduin in the Wadden Sea in mid-July 2023 (there is a breeding colony of Little Terns on the island, though 'our' bird appeared to be a non-breeder).

Nonetheless, as Kieran gleefully pointed out, his recovery of the adult increased the Ponds' own recovery rate by one third. To three.

It is undeniably a lot of work for three recoveries. Our wider ringing and monitoring effort is of course only one tiny part of the much larger data-gathering project but, even taken as a whole, and for all the tens of thousands of hours put in by dedicated ringers, surveyors and scientists, amateur and professional, the overall picture is far from complete. We can see that many, perhaps most, adults leave as soon as breeding is complete, and their young often go with them. And they can move quickly: one bird ringed in Essex was seen in Portugal six days later. But it is also the case that large flocks assemble in the Netherlands in late August, presumably to complete moult and further fatten up: it might seem obvious that UK birds should be part of these coastal gatherings, but there has been just one ringing recovery to support this. More surprisingly there's also been a British recovery at the other big European gathering site in the Venice Lagoons, but most, though not all, of the birds there seem to have come from the Adriatic and Dalmatian coasts – a maximum distance of 150km away – and those that haven't originate in

---

24 The North Lincolnshire colonies have been in steep decline since the 1990s and are currently virtually inactive. Comparative figures compiled by Liam Andrews suggest at least a partial relocation of the populations to Beacon Ponds.

*Little Terns seem to have a particular aversion to Kittiwakes, and
mob them relentlessly even as the season winds down.*

mainland northern Europe. Even their ultimate destination is uncertain: it
seems British Isles Little Terns winter on the West African coast in a relatively
small area off Mauritania, The Gambia, Sierra Leone and Guinea-Bissau (three
recoveries up to 2016), but some may find themselves further to the south-east
off Ghana and Côte d'Ivoire, where the majority of birds appear to come from
those northern European populations.

'Perhaps', 'often', 'presumably', 'seem', 'appear'…

For all that, our knowledge has expanded a great deal over the last 80 years. The entry on Little Tern migration in 1944's *Handbook of British Birds*, while noting that birds ringed in Suffolk and Yorkshire (presumably at Spurn) had been recovered on the French and Portuguese coasts, rather despairingly concludes that otherwise there is 'No evidence of passage to or from breeding grounds outside Brit. Is.'.

As the season continues to wind down, we begin to dismantle our equipment: the net fence on the beach is first to be disassembled and taken into storage, followed by the top strands of the main electric fence. Once the last tern has gone, the rest of the electric fence and the New Hut will be taken apart and stowed away. Tom, the longest-serving of our residential counters, will leave at the very end of the season. He has two more conservation gigs to see him through the winter, on Lundy Island and at Llanelli WWT, and we hope he'll return to us next year, perhaps as an assistant warden. Young as he is, his CV is already impressive.

James himself leaves a week earlier than he originally planned (taking the holiday entitlement he hasn't used earlier) in order to move house from one remote Scottish island to another. His contribution has been immense. Rob will revert to the general wardening role he had before the start of the season but, like then, will still find time to contribute to what needs to be done at the colony.

For the rest of us, we'll be into reviewing the season, analysing and interpreting the data we've gathered, writing short and long reports on it for partners and funders, and – of course – fundraising. Implicit in all that is planning for next year, and beyond.

I'm sometimes asked whether the Little Tern Project is 'worth it'. Five months of wardening, much of it on a 24/7 basis, sometimes as many as 20 volunteers, the erection and dismantling of electric fences and the rest of it, and the endless grant applications to fund the operation can seem like a disproportionate amount of energy to spend on what is, in national terms, a small colony (accounting for little more than 1 percent of the national breeding population) and one the long-term future of which is uncertain. And anyone who has spent any time in the huts watching the birds has sooner or later confronted the bitter thought that they seem to do little to help themselves: nesting in the open on low shingly beaches exposes them every day to predation from other birds and from mammals, and the risk of being wiped out at any point by bad weather, storms or high tides.

But it works.

Or at least, it works so long as there are enough colonies nationally to sustain the losses of one or two. The ruthless mathematics of natural evolution allow for it. The problem comes when there simply aren't enough colonies. And that, I'm afraid, is down to us, the human race.

'Human disturbance' doesn't just mean the direct deliberate or accidental interference of individuals at a colony: it also, and more seriously, means the more widespread destruction of habitat for human development and leisure activities. This makes the loss of any one colony all the more serious. The mathematics start to work against the birds. So yes, our own, positive brand of 'interference' makes the Project 'worth it'. We have to try to redress the imbalance we humans have ourselves created.

Maintaining and developing habitat is key. On the simplest level, this can mean providing the artificial hidey-holes for chicks or burying flowerpots full of sand in an otherwise too-shingly beach to form a readymade scrape (an experiment tried at Point of Ayre on the Isle of Man). It also means mainte-nance – removing invasive plants, for instance, something we ourselves are planning to try to get permission for before next season. It means antici-pating the needs of the birds as problems arise: in our case, the possibility that our individuals may move further south to their old stomping-grounds on the peninsula. On a bigger scale, it can mean restoring lost habitat, as at Minsmere in Suffolk where shingle is being added to islands to attract birds which were forced by erosion to stop using the site 15 years ago. At Horsey Island in Essex, sand and shingle from nearby channel-deepening work in Harwich Haven have been employed effectively to raise a beach that had become increasingly vulnerable to flooding because of rising sea levels, again to the benefit of terns and other shore-nesters. And on a grander scale yet, it can mean wholesale habitat creation: at RSPB Hodbarrow in South Cumbria a new island has been built in a lagoon, greatly increasing the space available for nesting Little Terns – and numbers have doubled (Common Terns have benefited as well, as have Eiders). And we have already seen how a disused airfield on Tiree has formed the basis of a thriving colony. There are other such projects around the country, too. What is notable about many of these initiatives is the positive use they have made of reclaimed materials, and in the case of Hodbarrow (a former iron mine) and Tiree, those materials are the products of less sympathetic historical interference in the landscape by humans. It's tempting to see a kind of natural, even poetic justice in this, but more importantly it points to a bigger truth: that not all human activity, not even that which is originally, even entirely, inimical to nature, need ultimately be entirely to nature's detriment. The limestone slag at Hodbarrow, the smashed concrete on Tiree, have made good habitat. I'm irresistibly reminded of our own landscape at the Ponds, of the human detritus, shattered brickwork and everlasting plastic aggregated here by nature itself but still, in some ways at least, to the benefit of the birds.

We might have begun closing the colony down, but – and this happened last year too – a lone bird is making us wait, a single fledgling that is not ready to go. So we hold off completing the wind-down and continue our monitoring duties, albeit on a reduced scale.

I've been looking forward to what is very likely going to be my last shift of the season – a morning session, starting at 6am, on the 24th. I walk up again from the south, parking on the edge of Kilnsea village. The sun is low over the horizon, the sea flat. A Swallow continues to take food into a nestful of young hidden in the shell of a Second World War gun emplacement, wrecked almost beyond recognition now but still massive as it continues to cling on after 80 years. It seems to have been a good breeding year for the Swallows: they've benefited from the hot summer and many have got two broods out. This one might be a third, though they will do well to fatten up enough to make it this late in the season: their fellows have been on their way south for a while now. There'll be hundreds more over the course of this morning, with pulses of House Martins and a few late Sand Martins thrown in. Half-a-dozen Swifts, too, though these birds are laggards: the great majority are long gone.

I walk up along the tide-line where the sand is still wet. The falling tide is as lazy as you'll ever see it, no more than the gentlest lapping. Arriving level with the southern end of the colony I cut across the beach, over the dunes and reach the New Hut.

I set up my scope and scan immediately for the last remaining chick, but it's not in its usual place and I can't see it anywhere else.

Still, the colony is full of birds, dominated by the white floes of hundreds of gulls, mainly Black-headed, gathering themselves after roosting to fly off to follow tractors in nearby fields. Nearly 50 Little Egrets are still dozing along the far shore, and a dozen hunched Herons stand sentry. The Shelducks and their five remaining ducklings are also napping fitfully on the end of a newly exposed spit. Five survivors out of twelve is not that bad a result for their parents. A large flock of Starlings gusts about; Linnets and Yellow Wagtails call from the edges of the dunes. Twenty more Wagtails fly over south, in several small groups. Most of the waders are away on the Humber, but there are still plenty to look at here: adults and this year's young are adding to the earlier non-breeders. Dunlin and Redshank dominate, but there are also scarcer visitors: a couple of Greenshank, a Ruff, a Curlew Sandpiper, two Little Stints and, scarcest of all, a delicate Red-necked Phalarope that's been around for a few days.[25]

Our fledgling could easily be invisible among that lot.

Three people approach from the north, with four dogs, all off-lead, but they turn back just before reaching the colony, perhaps because they saw me.

It ought to feel busy, but somehow it doesn't. For all the numbers of birds present – and there must be at least a thousand all told – the late August landscape of the colony feels drained of energy, listless, unsettled. Something is finishing. The Ponds are much reduced in size and edged in scummy, lurid green algae, the mud baked and bare after the hot dry summer. The grasses are withered and sere; much of the vegetation is grey or brown and desiccated. The saturated glaucous-blue of the Sea-holly has bleached almost to white. Sheets of low-growing Glasswort – Marsh Samphire – bring

---

25 Remarkably, the 'Phal' was to be joined by a second, then a third bird over the next few days. The species is officially designated as a 'rare visitor' in Roadhouse's *Birds of Spurn*. What constitutes 'rare' has increasingly become a moveable feast in recent years, but birds deemed particularly notable at the Ponds include White-rumped Sandpiper, Kentish Plover, Black Kite, Lesser Crested Tern, White-winged Black Tern, Audouin's Gull, Least Sandpiper, Terek Sandpiper and Greater Sand Plover. Europe's first Stilt Sandpiper was recorded just to the north in 1954.

a scabby purple-red to the reduced colour-palette. Even the Sea Asters are starting to tire.

Of course, we project our own moods on to the landscape, just as the landscape provokes moods in us: it's a mutual relationship. And even as a child I never much cared for the late summer and early autumn. Prosaically, it was probably to do with the ending of the holidays and the looming threat of the return to school, but it was – and is – also to do with a disconcerting sense of this being an in-between time, not quite one thing or the other. The dog-days of late August and early September, when the sun is still hot and the air can be at its stillest, are often described as sleepy, but if it is sleepy then it feels to me like a half-sleep, uneasy, fretful. Perhaps I'm just picking up on the birds' own twitchiness: those that are about to leave are showing a typical pre-migration restlessness in their behaviour (the Germans have a word for it: *Zugunruhe*). I get a whiff of that anxiety I felt when waiting for the terns to arrive in spring; 'what if they don't come? What would that mean?' Is this what it would feel like if they *hadn't* come? It's a foolish thought, I know, not least because the scene before me is packed with life. But it's enough to make me scan again for our last Little Tern of the season.

It has been favouring a new, small, pebbly island just out from the shore, like the Shelducks and the waders taking advantage of new land exposed by the falling water. The bird must have been close to leaving for a few days now: it's full-grown, with only a trace of down around its neck, and pretty active, scuttling back and forth and making short practice flights. Its juvenile plumage, however, means that even if it is still here this morning it will continue to be far from obvious, at least if it is keeping still.

A Sparrowhawk makes a run just north of me but catches nothing and swings away across the water to perch for a while on the gabions. A Kestrel – a female, not our problematic male – has been hovering over the marsh for some time.

Almost 30 Common Terns, a mixture of adults and young, followed by a few Sandwich Terns, come in-off, and settle nearby. I scan through them and continue scanning along the little island again.

Still nothing.

But the far edge has a bit of a lip in one or two places and the bird could easily be hunkered down out of sight behind one.

Two Hares lope easily across the sand without disturbing anything and slip into the dunes.

An adult Little Tern appears from nowhere, hovers briefly above the island then leaves. Was it a parent, encouraging its chick, invisible to me, to leave? Or was it confirming that its chick had already gone?

I've been here for an hour now, and I'm fairly confident the youngster *has* gone. If it has, it won't be back for two years, spending its time somewhere along the West African coast before returning to breed. And if it does manage to return, it may well not come back here to the Ponds other, perhaps, than as a passing visitor: Little Terns' loyalty to their natal sites is by no means guaranteed.[26]

I refuse the thought it might have succumbed to a predator overnight and return to scoping the Phalarope picking insects off the surface of the water at the north end.

But my attention is constantly drawn back to the little island in front of me. All around birds are leaving their roost as the tide turns and the day warms, but there seems little urgency around the colony even now. On the island itself, there remain just a couple of dozing Dunlin, a pert Ringed Plover…

…and a juvenile Little Tern, clearly ours and clearly very much still here.

Over the next couple of hours it barely moves. An adult bird – it seems obvious now it's a parent – visits half a dozen times, hovering above the juvenile and calling. It lands, briefly, twice. But it brings no fish, and the young bird seems unmoved by the attention it's getting. It's hard not to anthropomorphise: what I had started taking for lethargy, possibly even disease, on the part on the youngster now looks more like adolescent sulking, and the parent's calls seem to be taking on an increasingly exasperated if not defeated tone. Eventually, she(?) gives up and flies off purposefully to the south. There is no reaction from the juvenile, and none for the next half hour. I do a quick scan around

---

26 According to Cabot and Nisbet, birds bred in the south-east of the British Isles seem to relocate to the Netherlands; those bred further north, to Denmark and Sweden. The much smaller (approx. 400 pairs) Danish population seems more site-loyal, with almost 60 percent of birds returning to the same site (*Danish Bird Migration Atlas*).

to make sure I'm not missing anything important and when I get back to the tern it's vanished again.

I see nothing of it for the remaining hour-and-a-half of my shift, and neither do those who watch for the rest of the day and the day after.

It had flown. On my shift.

And I'd missed it.

AFTERWORD

Autumn 2023

We will almost certainly never know the fate of that last, stubborn fledgling. If it made it safely to its winter quarters, it will remain there for at least another year before attempting the return flight.

The photograph of the 2021 Little Tern pullus awaiting its ring is still on my desk, and with luck it might have made its way north to breed for the first time as an adult this year. We didn't see it here but, if it made it at all, it was at least as likely to fetch up at one of the Norfolk reserves, or on Teesside, or in the Netherlands. If it's been controlled, we'll know. Wherever it settled, it's not guaranteed that it bred or, if it did, whether it succeeded: first-time breeders don't always make the best choices.

Our post-season review of 2022 confirmed everyone's sense that we'd had a good year. Good weather (after a cold spring), relatively low levels of both predation and disturbance and a slight increase in the number of pairs of Little Terns attempting to breed led to 41 chicks fledging from 39 pairs (1.09). The absence of three-egg nests was possibly down to the late start to the season, with the poor spring perhaps meaning adults had less opportunity to get into peak breeding condition. It's also possible that they arrived in suboptimal condition after a hard wintering season. Ringed Plovers started badly but recovered reasonably well, while Avocets bred successfully for the first time since 2019. Oystercatchers, on the other hand, had a poor year, for reasons we still don't understand. The Kestrel problem accounted for at least ten tern chicks in the end, with our Sparrowhawk concentrating on adult Ringed Plovers. Plover eggs (a minimum of three nests' worth) were lost to Crows and (possibly) Hedgehog, and two nests were lost to high tides.

Our success was replicated and in several cases bettered by colonies around the country (we were in the top 15 for productivity. Not that anyone is counting…). I've already mentioned Gronant, the recovery at Chesil and the extraordinary results at Winterton. The big Irish colony at Kilcoole continued to expand, while Point of Ayre on the Isle of Man nearly quadrupled in size, its 34 nests turning out an excellent productivity of 1.94. Forvie in Scotland fledged 17 – its highest number since 2014. In all, at least 1,665 chicks fledged across the UK, Ireland and the Isle of Man.

And all this occurred as bird 'flu tore through other seabird species.

This season's birds, with or without our old pullus, showed up at the Ponds on a very similar schedule to last year: the odd passer-through was seen in late April, but the first settlers waited until 2nd May – just a day later than last year, and again on the back of a cold, wet spring. It was a particular relief to see them return, as news from their West African wintering areas had not been good: bird 'flu had broken out and decimated a number of wintering seabird populations, including those of the African Crested Terns and Caspian Terns.

There has been no news about the Little Terns there.

The new arrivals at the Ponds found their site somewhat improved. We managed to get consent to restore an area of habitat for them over the winter, clearing an area of about 0.6 hectares where wind-blown sand had been building up for a couple of years, allowing vegetation the opportunity to gain a foothold. We also expanded the perimeter fence to the north of the colony to encompass more land and take advantage of a natural break in the landscape. New strainer and metal support posts were put in for the electric fence. Three new wardens were appointed: Head Warden Jacob has wardened for us before while Tom, as hoped, returned as an assistant. Harry's a first-timer, as are our two residential volunteers, Eva and Erin. Rob is now operations manager for the Obs and still very much involved with the tern colony. We have a Leeds MSc student, Joe, gathering data for his dissertation as well as proving a very useful extra pair of volunteering eyes and hands. The New Hut was put up again at the south end, but the Old (northern) Hut finally gave up the unequal struggle and was demolished and carted away. Its successor was eventually put in place: an early-nesting Oystercatcher nearby held us up. Sea-buckthorn was cleared, litter picked. Our experiment for this season has been the provision

Habitat management in practice at the southern end of the site. Marram-grass and other invasive vegetation was removed and small, nascent dunes levelled.

of simple, raised nesting platforms for Oystercatchers, an idea borrowed from reserves in the Netherlands. Not only are Oystercatchers (like everything else, it sometimes feels) in need of our help as their populations fall, but by putting a couple of platforms at each end of the colony we hoped to develop our 'early warning system' to the benefit of all beach-nesters.

And this season has turned out even better.

We have had a record maximum count of 253 Little Terns at the colony. They won't be all ours, but as many as 140 of them are, and we have had at least 58 nests. The birds produced the (for us) staggering total of 107 successfully fledged young, giving a productivity of 1.84 chicks per pair. Neither figure is a record, but you have to go back a long way to beat the overall total, and productivity has exceed 2.0 only once – and that off single-figure pairs. We ringed 49 chicks, eight of them well grown enough also to have a darvic fitted.

In addition, 23 Ringed Plovers have fledged off 12 pairs. Six pairs of Avocets bred but suffered heavy predation later in the season, and we are uncertain how

Little Ringed Plover chick.

many, if any, chicks survived. Oystercatchers fledged from six nests (though two well-grown chicks were again mysteriously lost) and two lots of Black-headed Gulls successfully produced four chicks. A pair of Common Terns nested on the 'tern table' after a prolonged and uncertain courtship, but left it too late and failed. Despite the losses, overall these are excellent figures. And we had a new breeder, too: for the first time, a pair of Little Ringed Plovers, rarer cousins of our birds, successfully bred, raising two chicks.

Nationally, the Little Tern picture appears a little more mixed: populations of breeding pairs are up in some colonies, down in others. Chesil seems to be continuing to recover, the Irish colonies are going well as are those to the northeast of us, but south Norfolk seems to have quietened. We have seen that birds may move quite freely between colonies within seasons, but is it only because, through predation, weather, disturbance, food supply or failure to partner up, they *have* to? Might they also in some sense *choose* to, as part of a grander survival or expansion strategy? Some birds, perhaps most, seem to a greater or lesser extent to be site-faithful, but by no means all. Others might effectively be prospecting for new site opportunities, or 'testing' old ones: perhaps an unknown proportion of each generation is genetically programmed to pioneer. Might our colonies, often clustered into sub-units within the whole, be in the terns' eyes all sub-colonies within one great colony that is the entire British Isles, maybe with Dutch and Danish colonies too? And if so, what exactly is the evolutionary mechanism at work?

Speculation aside, it does seem the globe is still working.

There are several reasons why things have gone so well for us. Our area of reclaimed habitat was an immediate hit with the plovers and the Oystercatchers. In the case of the terns, they got down to breeding very quickly on arrival, including in the restored area. (Only one bird nested on the beach: it was lost, probably to Fox.) They also produced plenty of three-egg nests: it seems that, so long as the birds have fed well over the winter and arrive in good condition, the weather, provided it is not *too* bad, need not inhibit them. It also seems that the fishing here has been particularly good this year, with birds bringing in lots of good-sized prey. So while July was a very wet month – normally a major threat to survival rates, with eggs and chicks becoming chilled – the fact that plentiful fish were regularly available close to shore meant that parent birds spent minimal time away from the nest. All these factors may together account for the Beacon Ponds terns' breeding schedule 'catching up' with those other colonies which it has tended to lag behind in the past.

Predation has been light. Marcus's Kestrels bred again, but their eggs were predated. One pair, perhaps the same, regularly visited the area north of the Ponds and seemed to have taken up residence somewhere around Easington. They have, however, posed no real threat – less so than the Merlin which briefly resumed its former position as top gun among the smaller falcons. Foxes have been very regular, but penetrated the fencing only once; the same is true of the less frequent Badgers. Stoat was occasional. A large Grass Snake was seen, but we don't know if it has caused any damage; it is certainly capable, though any impact will have been light given its limited needs as a reptile. Little Owls returned as well, to a nearby farm, but showed no interest; it might be a different pair. We also made sure wardens were present on site in the early part of the season from before first light, prime hunting time for Crows; and we had some success in mitigating that threat. On the other hand, those losses of small plover and Avocet chicks were to some extent accounted for by Black-headed Gull – perhaps a single bird. (One, possibly the same bird, was also seeing harrying a tern into dropping fish, after the fashion of

A Black-headed Gull unexpectedly emerged as our top end-of-season predator.

a skua.) As is always the case with our defence against predators, it has been two-steps-forward-one-step-back.

And a new potential threat arose, a most extraordinary one: a Pine Marten was seen and photographed on the peninsula and in Kilnsea in the first weeks of the season. Its origin was initially unknown: it was thought that it might have escaped from captivity or even been deliberately released. The most intriguing possibility was that it was genuinely wild – and so it turned out to be. Photographs confirmed that it was an animal known as 'Two Spot', previously recorded in the Dalby Forest in North Yorkshire, some 60 miles away.

Although they are known to be expanding their range in England and Wales, having (probably) been restricted to the Scottish Highland forests for a hundred years, it is still hard to imagine a less likely place than Spurn for a Pine Marten to turn up. It is generally believed that one Pine Marten needs a range of 100 hectares of native woodland in its territory; there is nothing remotely approaching that anywhere here or even in our wider region. It may have been targeting the local Rabbits, but birds' eggs are also a favourite. So it might be – it is – a welcome sight elsewhere, but its presence nearby was not greeted with unqualified pleasure at the colony. On the other hand, the local Kestrel's eggs were predated by *something*... Fortunately, the Marten was never seen closer than a kilometre from the colony, and disappeared altogether mid-season.[27]

If predation has been light, then so has disturbance, with little significant to report with the exception of one incident: another 'first', and an unwelcome one at that. A carload of seven drunk men careered up the beach at 1am, perilously close to both plover and tern nests. The police were informed. We despair.

But the spectre of bird 'flu continues to haunt us. We have again had no confirmed cases in the colony in 2023, but major early-season outbreaks in inland Black-headed Gull and Common Tern colonies showed that the problem was far from over. Conservationists were better prepared this season: the National Trust, for example, decided pre-emptively to close the Farne Islands to day-trippers again: visitors have not been able to go ashore but could only circumnavigate in boats. Last season wardens had the desperate job of disposing of 6,000 dead birds of several species (though no Little Terns: they don't breed there). But this year at least one Little Tern was confirmed as having the disease at the Kilcoole colony in Ireland, and it was also reported that at Long Nanny in Northumberland, a reserve that escaped the virus last year, several Little Terns were suspected to have fallen victim to the disease. And a larger tragedy is unfolding there: what is the biggest mainland breeding colony of Arctic Terns has lost no fewer than 800 chicks to bird 'flu. Elsewhere, Forvie has lost at least 200 Sandwich Terns.

All any of us could do was what we did last year; perhaps a little quicker, perhaps a little better. There has been nothing new to try, nothing that can truly combat the virus. There are vaccines, but their efficacy is uncertain, not least because they might mask symptoms without dealing with the underlying disease – and how do you deliver them to wild bird populations anyway?

It is hard to over-estimate the continuing catastrophe for many of our breeding seabirds. Since its initial appearance in the Northern Isles, where it devastated Great Skua and Gannet colonies, bird 'flu has spread through the UK countries.

---

27 Remarkably, a Grey Squirrel – another potential egg-thief – was also seen in the colony. Not the first time (see note 21, above) one has been seen here, and not as wholly improbable as the Marten, but still very unusual.

With the Great Skua there is concern it might prove the trigger for an extinction-level event. Last season the world's largest Gannet colony on Bass Rock in the Firth of Forth collapsed. The disease spread south as far as East Anglia and Cornwall, with Sandwich Terns being particularly hard-hit – but also Black-headed Gulls, auks (including Puffins), Kittiwakes and other gulls, Shags… The numbers are horrific, and cannot easily take into account how many more fatalities occurred unseen, out to sea, nor how far our roving seabirds might go on to deliver the virus: its presence has now been detected in the Antarctic. Nor has the disease confined itself to seabirds, though they bore the brunt: the BTO had confirmed cases in 61 wild bird species by October 2022, and these on top of the millions of domestic fowl (3.5 million in Britain alone) necessarily culled. There have also been cases of the virus jumping to other species, including Otters, Foxes and Seals.

There is some good news, though: according to researchers from the RSPB and Scottish universities, some Gannets, and a few other species, have shown signs of being able to survive the virus. The tell-tale sign is a change of iris colour from (in the case of Gannets) the usual ice-blue to a uniform or mottled black. The majority of birds with black eyes that have been tested have been revealed to have had, but survived, the virus. And the disease was detected in four species on the Isle of May in Scotland, yet did little real damage: Puffins had a successful breeding season there. Little Terns seemed safe: there were no confirmed reports of mortality from any of the UK colonies last season. But summer's news from Long Nanny was unsettling. It remains possible that even if they are not immune, Little Terns might have some resistance to the disease: or like the Gannets, some Little Terns might survive infection and develop immunity. But it is equally possible that a new strain of the virus might have evolved to take advantage of a new host species. The problem is that the UK population is already small and, even with help, is maintaining only a limited presence.

And this really is the issue.

Those who argue that conservationists are being alarmist about the impact of bird 'flu, that it has been present in the ecosystem for many years (it has, including in the Spurn area) do not seem to allow for the fact that viruses are by their very nature highly adaptive and constantly produce new strains. That is what it is to be a virus, as Covid has shown us only too well. The great majority of new strains go, from the virus's point of view as it were, to waste, but others do not. It's also argued, rather airily, that nature's powers of recovery have been proved to be remarkable, and so they are – given a chance. But are we giving nature a chance? Seabird populations around the UK are already under severe pressure: pollution, overfishing and loss of habitat play serious parts, but the greatest problem with the most widespread impact is that of climate change. Rising sea levels threaten low-lying breeding grounds. Warming seas interfere with the breeding cycles of prey items and push cold-water species farther to the north, or down to the cooler depths, beyond the reach of many birds. (Recent research finds that surface temperatures in the North Sea were several degrees above normal in 2023, though this seems not to have immediately affected our prey populations.) Fishing can also become more difficult as a warmer climate produces cloudier seas. Seabirds are long-lived, too, and many species are slow to breed and produce low numbers of young, meaning populations, even given a fair wind, take a long time to recover. An event like the current outbreak could be the final straw. Perhaps it already has proved so for the Great Skua: we don't know yet, and we won't for a while.

And of course this is only one part of a much bigger picture of relentless loss and continuing decline in the natural world, enabled by political compromise, by promises broken and deadlines missed, by bad faith from too many governments (including our own) and the big fossil fuel companies. Sometimes it feels that for all the hard work, energy and bravery of conservationists, the war is already lost. I have talked to young conservationists who take it for granted that it is – but still commit themselves wholeheartedly to voluntary work or to professional careers in the area. It is as well that they do: none of us can afford the indulgence of despair.

Does it matter if Yorkshire loses its Little Terns? Or even if the British Isles do? I think it does. If the pressures at work on these birds were purely local or national, it might not: as I've said, they are doing relatively well across their international range. But the problems are global. Just as the loss of any UK or Irish colony, even a small one, increases the pressure on the others, so the loss of all the British and Irish colonies would impact across Europe. Where does it stop?

And in any case, the threat to our Little Terns, and to the natural world generally, is a threat to us, physically, mentally and spiritually.

I think back to my first encounter with Beacon Ponds 20 years ago, how empty and desolate the area seemed to me. As I write now the colony is as devoid of terns as it was then: there is just the odd straggler hurrying through on its way south. But my perception of it has changed entirely. A few years of immersion in the place and its many moods, of observing it closely – what Gilbert White of Selborne, one of the earliest and undoubtedly the greatest of our parson-naturalists, called "watching narrowly" – has enriched my understanding immeasurably. It is not just that I know more about the terns and about the other birds which live and breed here, but I've learned about the mammals and plants and even the invertebrates in the water too: their relationships, their interdependence, their community. The human community too, its history and particular culture. And our attention as watchers and guardians has infused the character of the place as well, freighted it with meaning, thickened the landscape through anecdote, story, character, relationship... through being human. Like White, we develop a holistic, empathetic sense of the place we are in and the other lives we share it with. Lose the Little Tern from our beaches and shorelines, or the Curlew from the moor, the hawk from the sky, the Blue Tit from the garden – even the hybrid duck from the park – and another thread is pulled out, the landscape thins, our sense of place is impoverished and we with it.

In the summer of 2022, the poet Kathleen Jamie made her annual birthday trip to the Scottish isles. This time, her visit carried an extra purpose: she was seeking solace – a sense of life going on, of "old familiar comforts", following her father's recent death. Instead, she saw conservationists in hazmat suits filling bin bags with the bodies of dead seabirds. In fact, "it's been a while", she says, "since we have been able to turn to the natural world for reassurance, to map the arc of an individual life against the eternal cycle of the seasons, the birds, the hills…".

It is not a happy insight, but I fear it is a true and accurate one.

We were spared that catastrophe at Beacon Ponds in the 2022 and 2023 seasons, and we continue to draw on "old comforts": the pleasures and rewards that come from working together for our Little Terns. In the 2022 season, with these people, in this special, ever-changing place, whatever was happening in the wider war, a battle was won. We won again this season. But like the terns themselves, we are clinging to the edge.

APPENDIX

The Season in (some) Statistics

To reproduce all the documents, tables and spreadsheets that came out of the 2022 season would require another book – and a rather dry, not to say unreadable one, at that. Here then is a selection, much abbreviated, from the range of data we collect (that range also includes, for example, information relating to the weather, the tides and the performance of the electric fence).

All data are destined, as appropriate, for SBOT, SHCS, RSPB (via the National Little Tern Project), EA, NE, BTO, HNP.

Tables have been compiled by James Wilson and Mike Coverdale.

1. Little Terns

Date of first arrival: 1 May 2022

Date of last departure: 24 August 2022[28]

---

28 Of resident birds; the last records of passage birds were of two on 22 September and a very late bird flying south on 9 October. The latest ever was recorded one November.

Table 1: Maximum daily counts

Day/month	May	June	July	August
1	4	76	84	112
2	3	57	51	87
3	5	10	100	132
4	8	41	92	102
5	8	64	75	110
6	18	60	99	49
7	16	74	98	46
8	11	44	93	56
9	10	48	90	59
10	31	67	88	46
11	25	77	125	42
12	8	96	64	60
13	24	82	106	45
14	24	88	72	25
15	17	115	102	45
16	40	82	66	32
17	41	90	102	21
18	44	87	82	10
19	44	98	52	12
20	61	52	85	2
21	54	54	65	2
22	74	42	81	4
23	62	73	70	0
24	28	68	79	2
25	45	78	74	3
26	57	82	117	0
27	49	81	112	1
28	74	69	102	-
29	54	80	205	-
30	87	89	121	-
31	74		74	-

Table 2: Proportion of different clutch sizes in Little Tern nests found during colony walk-throughs

Eggs per nest	Number of nests	Total number of eggs
1	8	8
2	19	38
3	0	0

Breeding Productivity: Forty-one chicks fledged from 39 pairs, giving an annual rate of chicks fledged per pair of 1.05. The rolling five-year average is 0.88 chicks fledged per pair, above the sustainability target of 0.75.

Table 3: Historical Little Tern productivity

Year	Number of pairs	Number of young fledged	Productivity chicks fledged per pair	Rolling 5-year average productivity chicks fledged per pair	Rolling 5-year average number of pairs	Rolling 5-year average chicks fledged per annum
1977	5	2	0.4			
1978	4	0	0			
1979	5	0	0			
1980	6	0	0			
1981	5	0	0	0.08	5	0.4
1982	4	8	2	0.33	4.8	1.6
1983	6	15	2.5	0.88	5.2	4.6
1984	8	23	2.88	1.59	5.8	9.2
1985	11	8	0.73	1.59	6.8	10.8
1986	22	5	0.23	1.16	10.2	11.8
1987	2	0	0	1.04	9.8	10.2
1988	3	4	1.33	0.87	9.2	8
1989	25	1	0.04	0.29	12.6	3.6
1990	31	29	0.94	0.47	16.6	7.8
1991	20	0	0	0.42	16.2	6.8
1992	34	11	0.32	0.4	22.6	9
1993	62	20	0.32	0.35	34.4	12.2

Year	Number of pairs	Number of young fledged	Productivity chicks fledged per pair	Rolling 5-year average productivity chicks fledged per pair	Rolling 5-year average number of pairs	Rolling 5-year average chicks fledged per annum
1994	65	29	0.45	0.42	42.4	17.8
1995	71	4	0.06	0.25	50.4	12.8
1996	49	31	0.63	0.34	56.2	19
1997	42	2	0.05	0.3	57.8	17.2
1998	41	42	1.02	0.4	53.6	21.6
1999	54	45	0.83	0.48	51.4	24.8
2000	49	9	0.18	0.55	47	25.8
2001	44	3	0.07	0.44	46	20.2
2002	34	32	0.94	0.59	44.4	26.2
2003	30	0	0	0.42	42.2	17.8
2004	34	0	0	0.23	38.2	8.8
2005	27	28	1.04	0.37	33.8	12.6
2006	51	4	0.08	0.36	35.2	12.8
2007	27	0	0	0.19	33.8	6.4
2008	30	1	0.03	0.2	33.8	6.6
2009	26	0	0	0.2	32.2	6.6
2010	11	1	0.09	0.04	29	1.2
2011	25	19	0.76	0.18	23.8	4.2
2012	23	11	0.48	0.28	23	6.4
2013	36	40	1.11	0.59	24.2	14.2
2014	45	60	1.33	0.94	28	26.2
2015	37	30	0.81	0.96	33.2	32
2016	6	5	0.83	0.99	29.4	29.2
2017	49	14	0.29	0.86	34.6	29.8
2018	33	4	0.12	0.66	34	22.6
2019	25	39	1.56	0.61	30	18.4
2020	27	42	1.56	0.74	28	20.8
2021	38	17	0.45	0.67	34.4	23.2
2022	**39**	**41**	**1.05**	**0.88**	**32.4**	**28.6**

2. Ringed Plovers

Table 4: Breeding statistics

No. of pairs	No. of chicks	No. successfully fledged	Overall no. of nesting attempts	Maximum no. of concurrent active nests
16	30–40	15–20	24	12

Table 5: Hatching success of Ringed Plover nests by nest location

Nest Location	No. of nests	Hatched	Not Hatched	Chicks hatched no.	Proportion of nests hatched (%)	Chicks hatched per pair
Main fence	14	5	9	17	35.71	1.21
Beach fence	4	2	2	4	50.00	1.00
Outside of fence	6	2	4	5	33.33	0.83

3. Oystercatchers

Table 6: The last five years

Year	No. of pairs	No. of chicks fledged	No. of successful nests	Productivity, chicks fledged per pair	Rolling 5-year average productivity	% nests successful to fledging per year
2018	5	3	?	0.6	0.7	?
2019	5	4	2	0.8	0.7	40
2020	4	5	4	1.3	0.7	80
2021	4	1	1	0.3	0.7	20
2022	7	2	1	0.3	0.6	20

4. Birds found dead

Table 7: Corpses found (no confirmed bird 'flu; numbers never reached the Defra reporting bar)

Sandwich Tern	3
Common Tern	1
Little Tern	1
Cormorant	1
Kittiwake	2
Gannet	2
Guillemot	1
Oystercatcher (chicks)	3

5. Disturbance

Table 8: Summary of recreational disturbance data

Activity	None 1	Slight 2	Moderate 3	High 4	Extreme 5	Total
Birdwatching	2	4	4	2		12
Cycling	-	-	1	-	-	1
Walking	2	6	11	2	1	22
Dog off-lead	8	6	15	3	3	35
Dog on-lead	2	2	9		1	14
Fossil hunting		2	2			4
Horse riding		-	4	-	1	5
Off-roading	3	1	1			5
Low-flying aircraft – fixed wing	-		-	1	1	2
Low-flying aircraft – helicopter			-	-	1	1
Microlight	-	-	-	-	-	-
Paramotor	-	-	-	-	-	-
Bomb disposal	1					1
Photography	-	4	1			5
Angling	1	1	2			4
Golf	1					1

6. **Wardening**

Table 9: Summary of wardening hours

	Number of hours	Average hours per shift	Number of shifts
Total warden hours	2,158	6:00	356
Employed wardens	1,221	6:45	178
Residential volunteers	690	5:15	121
Local/day volunteers	247	4:15	57
Total for volunteers	937	5:30	178

IMAGE CREDITS

All the images in this book are reproduced by kind permission of their creators.

Photographs are by Richard Boon (pp. vi, xviii, 2, 8, 15, 25, 26, 101), David Constantine (frontispiece, pp. 4, 11, 33, 36, 38, 41, 44, 55, 58, 60, 62, 63, 68, 79, 86, 92, 96), Robert Hunton (p. 76 bottom), Michael Pilsworth (p. 113), Jackson Sage (p. 76 top), Thomas Willoughby (pp. 45, 49, 77, 80, 82, 90, 98, 116), Thomas Wright (pp. 114, 120), Georgia French (p. 16), Bethan Clyne (p. 103) and Alice Fox (p. 29).

The sketch on p. 64 is by James Wilson.

The map on p. xvii is by Richard Boon.

The portrait of John Cordeaux on p. 6 is freely available in the public domain.

Data relating to the colony and its history are included by kind permission of the Science and Research Committee of the Spurn Bird Observatory Trust and David Constantine.

Data drawn from the websites of the British Trust for Ornithology, the Joint Nature Conservation Committee, the Little Tern Project, BirdGuides and Rare Bird Alert are freely available in the public domain.

SELECT BIBLIOGRAPHY

The principal sources for this book have been Cabot and Nesbitt's 2013 monograph *Terns* (London: Collins); though a decade old, this remains the key text for anyone interested in the British terns. The late Andy Roadhouse's monumental achievement *The Birds of Spurn* (Kilnsea: Spurn Bird Observatory) was published in 2016, but is digitally updated annually. John Mather's *The Birds of Yorkshire* (London: Croom Helm), despite being published as long ago as 1986, is still required reading, and includes useful biographical and historical material as well as site accounts and a systematic list. Dr Jan Crowther's 2006 *The People Along the Sand. The Spurn Peninsular and Kilnsea: A History, 1800–2000* (Andover: Phillimore) is an indispensable account of the history and culture of the Spurn area.

The annual reports of the Beacon Ponds/Easington Lagoons Little Tern Project, 2000–present, various authors, have been invaluable. They are not yet publicly available, but it is hoped that they may be linked to the SBOT website soon (https://spurnbirdobservatory.co.uk). That site is also more than a little useful in its own right. The written archive of the Observatory, dating back to 1946, is to be found in the East Riding Archives at the Treasure House in Beverley and may be accessed on application to the Head Archivist. Spurn Bird Observatory *Reports* and later *Spurn Wildlife*, published by the YNU and the Observatory under various editorships, provide useful information from the 1970s to the present. Archive material from SHCS, as held in the private collection of David Constantine, though incomplete, takes the story back into the 1990s via newsletters and other material.

The website of the National Little Tern Project is to be found at https://littleternproject.org.uk. Birdguides.com was an essential means of keeping in touch with the developing national context of the 2022 season, as was Rare Bird Alert (https://rarebirdalert.co.uk). Both were accessed frequently. The Joint Nature Conservation Committee's website (https://jncc.gov.uk/our-work/little-tern-sternula-albifrons) and the British Trust for Ornithology's site (https://www.bto.org) were also useful background resources.

Other works consulted include:

Ackerman, Jennifer (2016) *The Genius of Birds.* London: Corsair.

Armstrong, Patrick (2000) *The English Parson-Naturalist: A Companionship Between Science and Religion.* Leominster: Gracewing.

Babcock, Michael and Booth, Viv (2020) *Diversionary Feeding Kestrels: Tern Conservation Best Practice.* Produced for 'Improving the conservation prospects of the priority species roseate tern throughout its range in the UK and Ireland'. LIFE14 NAT/UK/00394; live document last updated October 2020. http://roseatetern.org/diversionary-feeding.html# Accessed July 2022.

Bevan, Will (2022) 'A Bumper Year for the Little Terns of East Norfolk and North Suffolk'. *Life on the Edge Project.* https://www.projectlote.life.news (accessed 28 August 2022).

Bewick, Thomas (2010) *Bewick's British Birds.* London: Arcturus. First published 1826.

Bhatia, N., and Franco, A. (2015). Saline Lagoon Survey of The Lagoons Site of Special Scientific Interest (SSSI) – Easington Lagoons, North Humber. A report of the Institute of Estuarine and Coastal Studies (IECS) to Natural England. Final Report No. YBB253-F-2015.

Birkhead Tim (2012) *Bird Sense: What It's Like to Be a Bird.* London: Bloomsbury. https://doi.org/10.1016/S0262-4079(13)61893-X

Blow, Oliver (2021) 'Factors Affecting Nest Site Selection and Predator Visitation of Little Terns (*Sternula albifrons*) at Beacon Ponds Reserve'. Unpublished University of Leeds MSc dissertation.

Chandler, Richard and Wilds, Claudia (1994) 'Little, Least and Saunders's Terns'. *British Birds* 87: 60–66. Also at https://britishbirds.co.uk/content/little-least-and-saunders-little-terns. Accessed May 2022.

Cocker, Mark and Mabey, Richard (2005) *Birds Britannica.* London: Chatto and Windus.

Cordeaux, John (1872) *Birds of the Humber District.* London: van Voorst. https://doi.org/10.5962/bhl.title.29880

Davies, Charlotte (2011) 'Temporal and age-specific limitations upon reproductive success of Little Terns'. Unpublished University of Leeds MSc dissertation.

Davies, Stephen (1981) Development and Behaviour of Little Tern chicks. *British Birds* 74: 291–98. Also at https://britishbirds.co.uk/content/development-and-behaviour-little-tern-chicks (Accessed June 2022).

Easington Parish Council website http://www.easingtonparishcouncil.co.uk/history.aspx (Accessed 3 November 2022.).

Fox, Alice (2012) *Textures of Spurn.* Sunk Island: SC Publications. See also https://www.alicefox.co.uk

Gordon, Helen (2021) *Notes from Deep Time: A Journey Through Our Past and Future Worlds.* London: Profile Books Ltd.

Griffin, Joseph (2023) 'The influence of vegetation on nest site selection and rest in breeding shorebirds. Evidence for management outside the breeding season?' Unpublished University of Leeds MSc dissertation.

Jackson, Christine E. (1968) *Bird Names*. London: H.F. and G. Witherby Ltd.

Jamie, Kathleen (2022) 'Diary'. *London Review of Books* 44, 16 (18 August 2022): 37.

Johns, the Rev C. A. (1862) *British Birds in their Haunts*, edited, revised and annotated by J. A. Owen, 1910. https://doi.org/10.5962/bhl.title.13591

Kuiken, Thijs and Cromie, Ruth (2022) 'Protect wildlife from livestock diseases' (editorial) *Science* 378, 6615 (6 October): 5. https://doi.org/10.1126/science.adf0956

Mabey, Richard (2005) *Nature Cure*. London: Chatto and Windus

Morris, Francis Orpen (1862–1867) *A History of British Birds*. 6 vols. London: Groombridge. Available at https://www.biodiversitylibrary.org

Moxom, D.J. and Burden, R.F. (2004) 'The Recent History and Monitoring of Little Tern *Sterna albifrons* and Common Tern *Sterna hirundo* on Chesil Bank, Dorset, 1974–1999'. *Dorset Proceedings* 126: 63–84.

Newton, Ian (2010) *Bird Migration*. London: HarperCollins.

Owens, Susan (2020) *Spirit of Place: Artists, Writers and the British Landscape*. London: Thames & Hudson Ltd.

Pashby, Brian S. (1988) *A List of the Birds of Spurn 1946–1985*. Kilnsea: Spurn Bird Observatory.

Potter, Gwen (2023) 'I helped pick up 6,000 dead birds last summer. This is what I learned about the horrors of bird flu', *Guardian*, 4 January 2023.

Richardson, Jane *et al.* (2014) 'Earlier Prehistoric Activity and a Later Iron Age and Roman Field System at Beacon Lagoons, Kilnsea, East Riding of Yorkshire', *Yorkshire Archaeological Journal* 86 (1): 3–32. https://doi.org/10.1179/008442 7614Z.00000000043

Robinson, R.A. (2005) BirdFacts: Profiles of birds occurring in Britain & Ireland. BTO, Thetford. http://www.bto.org/birdfacts (Accessed on 23 May 2022)

Smart, Jennifer and Amar, Arjun (2018) 'Diversionary feeding as a means of reducing raptor predation at seabird breeding colonies', Journal for Nature Conservation. https://doi.org/10.1016/j.jnc.2018.09.003

Snow, D.W. and Perrins, C.M. *et al.* (1998) *The Birds of the Western Palearctic vol. 1 Non-Passerines*, concise edn. Oxford: Oxford University Press.

Wells, Tom (2021) *Big Big Sky*. London: Nick Hern Books. https://doi.org/10.5040/9781784606626.00000002

Weston, Phoebe (2020) 'The scale is hard to grasp: avian flu wreaks devastation on seabirds', *Guardian*, 20 July.

———— (2022) 'Falling birds and dehydrated hedgehogs: heatwave takes its toll on UK wildlfe', *Guardian*, 25 July.

Wilson, Andrew and Slack, Russell (1996) *Rare and Scarce Birds in Yorkshire*. Guildford: Biddles Ltd.

Wingfield Gibbons, David, Reid James B. and Chapman, Robert A., comps. (1993) *The New Atlas of Breeding Birds in Britain and Ireland: 1988–1991*. London: T. & A.D. Poyser.

Witherby, H.F. *et al.* (1944) *The Handbook of British Birds*, vol. 5. London: H.F. and G. Witherby Ltd.

INDEX

Note: Page numbers in *italics* refer to illustrations.

Pallas, Simon Peter 3
Peregrine 45, 54, 66, 90, 95
Piggott, Arthur 11
Pine Marten 116
predation 14, 53, 81, 85, 90, 105, 113–15, 117
predators 40, 42, 66, 78, 82, 95
Public Space Protection Orders
   (PSPOs) 43

Ramsar Convention 16
recreational disturbance data 127
Ringed Plover 3, 21, 23, 39
   breeding statistics 126
   cages 40
   chicks 90, 91, 93, 114
   hatching success of 126
ringing breeding birds 96
rings 97–8
   coloured 98
   darvic 98–9, 98
   numbered metal 98, 102
Roadhouse, Andy 97
Romano-British field system 4
Roseate Tern xii, 3

sand dunes 4
Sanderling 46, 57
Sandwich Tern 72–3, 108
seabird populations 119
Seabirds Preservation Act 1869 6
Sea-buckthorn 22
sea-swallows 46–7
Second World War 7–8
   bomb disposal 42
   gun emplacement 106

sewage discharge 28
Shelduck 66, 82, 90
Site of Special Scientific Interest
   (SSSI) 16
South Holderness Countryside Society
   (SHCS) 11
Sparrowhawk 90, 93, 108
Special Area of Conservation
   (SAC) 16
Special Protection Area (SPA) 16
spring-trap technique 97
Spurn Bird Observatory Trust (SBOT) 11
   Mike Coverdale leadership 13
Spurn Heritage Coast Project 11
Stoat 35, 95, 115
Sykes, Christopher 6

Taylor-Bruce, Jake 39
thermal-imaging telescope
   75–7, 76
Turnstone 46, 54

vandalism 12, 60

watcher, appointment of 7
White, Gilbert 120
wild bird, holding 99
Wildlife and Countryside Act 1981
   16–17
Wildlife Trusts 14
Willughby, Francis 31

Yorkshire Naturalists Union (YNU) 7–8
   role of 11
Yorkshire Wildlife Trust (YWT) 11, 50